THE MOST DIFFICULT GEOMETRIC PUZZLES

Published in 2020 by Welbeck Non-Fiction
An imprint of Welbeck Publishing Group
20 Mortimer Street
London W1T 3JW

Editorial: Georgia Goodall
Design Manager: Stephen Cary

A CIP catalogue for this book is available from the British Library.

ISBN: 978-1-78739-632-6

Printed in Dubai

10 9 8 7 6 5 4 3 2 1

THE MOST DIFFICULT GEOMETRIC PUZZLES

Tricky puzzles to challenge every angle of your spatial skills

Graham Jones

CONTENTS

INTRODUCTION

Welcome to what is not a geometry textbook. What you have here is something to get your mind in good shape, using shapes. It's a form of entertainment which should in part challenge, surprise, make you smile and provide a different perspective on things.

We all know the names of the great Greek geometers – Pythagoras, Plato, Archimedes, Euclid foremost among them – and the pleasure (or pain) they caused us at school. This isn't an exam, however, so you can breathe easy. You don't need to remember your theories from school to solve the challenges here, you just need to be prepared to think a little. And, a tip for you – sometimes you'll have to think outside the box.

Having said that, there are probably two things we all remember, or half-remember, about geometry. The first is Pythagoras' theorem ($a^2 + b^2 = c^2$) for right-angled triangles. The second is the story of Archimedes having a bath, displacing water, and running naked down the street shouting, 'Eureka!'. It all ended sadly for a goldsmith who had been commissioned to make a crown out of pure gold for King Hiero II.

The first thing to know about geometry and what makes it so good is that it's really useful. Geometry (literally 'earth measurement') began as a way to solve practical problems in surveying, building and astronomy, and the first written documentation of theories goes back an amazing 5,000 years to Babylonia and the Indus Valley. So, you're on a well-worn track, following in the footsteps of countless mathematicians.

The twenty different puzzle types inside won't solve astronomical questions, but they've been designed to exercise all sorts of mental skills – logic and reasoning, spatial relations, 2D and 3D thinking, arithmetic and lateral thinking. We've got some classic puzzle types, including matchsticks and tangram puzzles, and some new creations that you'll find along the way.

There are some things that link them all, though, including line and shape. Some will rely on your ability to think in 3D, some will need you to sit back, think and be inspired. And if you want to emulate Archimedes, you can even do them while in the bath. Bonus points for shouts of 'Eureka!'.

Another key characteristic that you should notice as you work your way through is that maths in general, and geometry in particular, is about relationships: how things work in relative terms and how things combine to form something new. So, some life lessons from geometry, as well as practical problem solving.

Best of all, though, we've got the answers for you (told you it's not an exam!), so if you get stumped, move on to something else and then go back. And then if you do get truly stuck, have a sneaky peak.

We'll let you at the puzzles now – they're of all shapes and sizes, for brains of all shapes and sizes. Wherever you fit in that spectrum, shape up and have fun.

– **Graham Jones**

PUZZLES

TANGRAM

The seven different shapes provided can be combined to form the outlined duck. How?

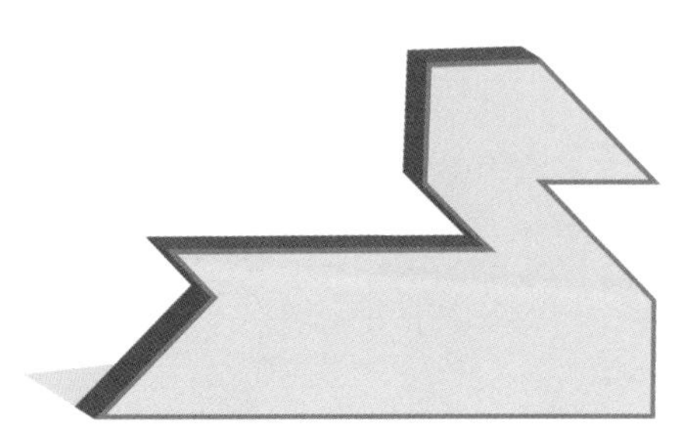

SOLUTION page 112

The seven different shapes provided can be combined to form the outlined dancer. How?

SOLUTION page 118

SQUARE DEAL

Split the shape into two pieces that can be arranged to form a 5x5 square.

SOLUTION page 112

04 Divide this shape into two equal halves, marking the point of rotation.

SOLUTION page 119

MATCHSTICKS

Move two matches to make there be only four squares.
Each match must form part of a square, and no matches can overlap.

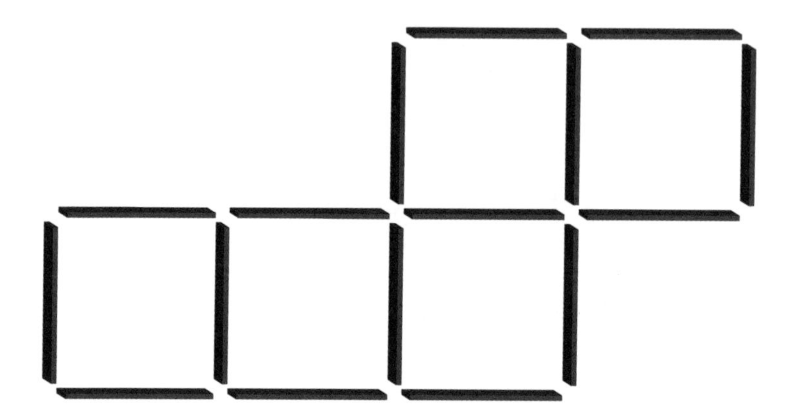

SOLUTION page 119

A) Move three matches to make five triangles.
Each match must form part of a triangle, and no matches can overlap.
B) Move four matches to make three triangles.
Each match must form part of a triangle, and no matches can overlap.

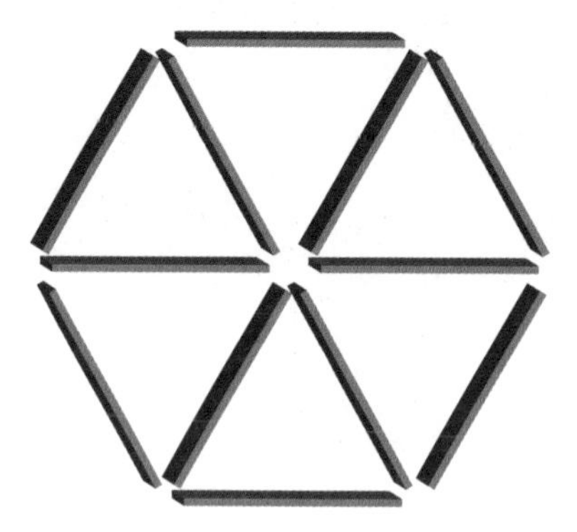

SOLUTION page 112

ADDING LINES

A) Where can you draw one straight line in this pattern to create four triangles?
B) Where can you draw one straight line in this pattern to create eight triangles?

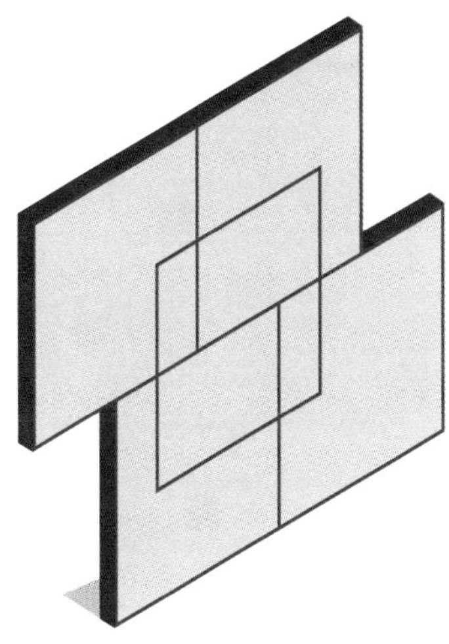

SOLUTION page 113

Add two straight lines to the star shape to create 13 more triangles in the shape.

SOLUTION page 119

LINE DRAWING

09 Use five lines, running from one side to another, to divide the rectangle into thirteen areas. Each area must contain one square. Areas cannot contain the same colour square as any adjacent area, even diagonally.

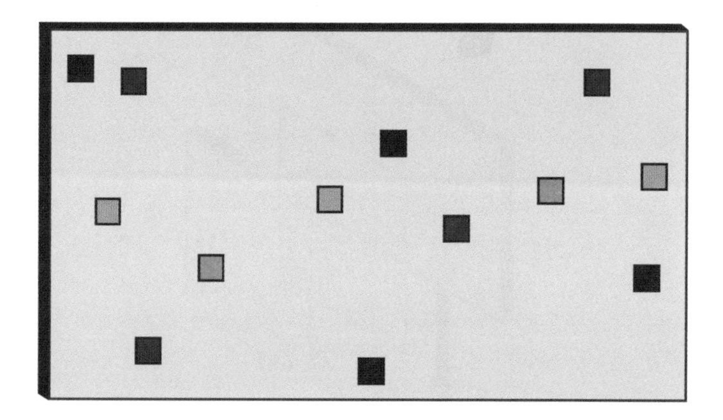

SOLUTION page 120

10 Use five lines, running from one side to another, to divide the ellipse into sixteen areas. Each area must contain one dot.

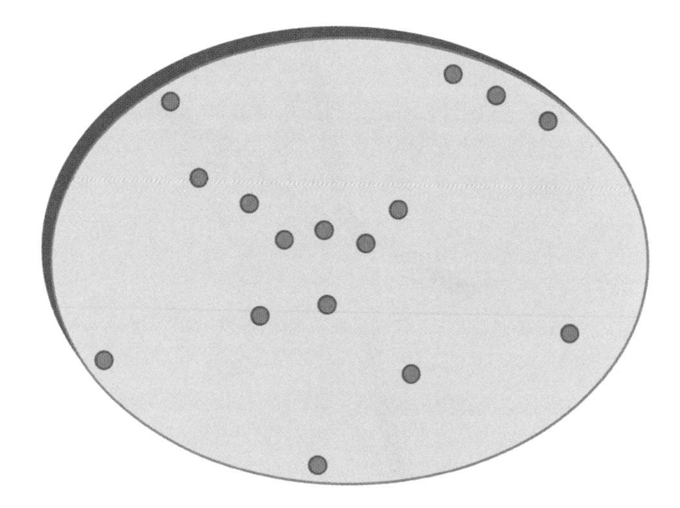

SOLUTION page 113

CONGRUENT SHAPES

11 Split the grid into two congruent shapes, identical when rotated and/or flipped.

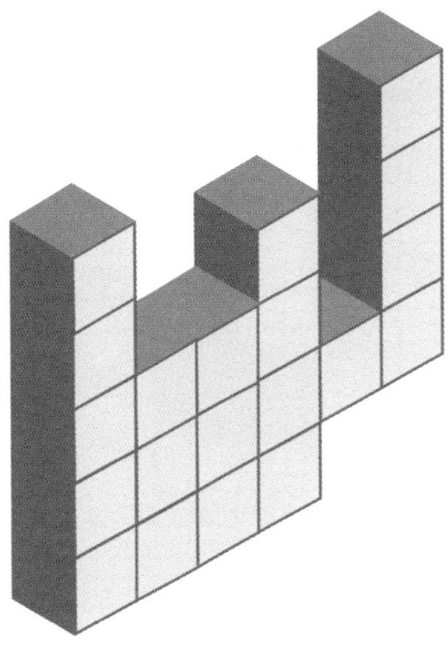

SOLUTION page 118

12 Split the grid into two congruent shapes, identical when rotated and/or flipped.

SOLUTION page 113

XOXO

13 Divide the grid into four congruent shapes, with the point of rotation at the centre of the grid. We have included four blue and four red squares as clues – no shape contains more than one of each.

SOLUTION page 120

14 Divide the grid into four congruent shapes, with the point of rotation at the centre of the grid. We have included four blue and four red squares as clues – no shape contains more than one of each.

SOLUTION page 114

OVERLAID

15 What is the smallest number of overlaid paper squares that can be used to create this pattern? In what order were they laid?

SOLUTION page 114

16 What is the smallest number of overlaid paper squares that can be used to create this pattern? In what order were they laid?

SOLUTION page 126

OVERVIEW

17 A 2x2x2 cube is made of eight differently coloured cubes of the same size. One of the cubes has been removed. Five of the six faces are shown, as seen from above. A thicker border indicates that this cube is on the bottom row. When looking at the sixth and final face, as seen from above, what do you see?

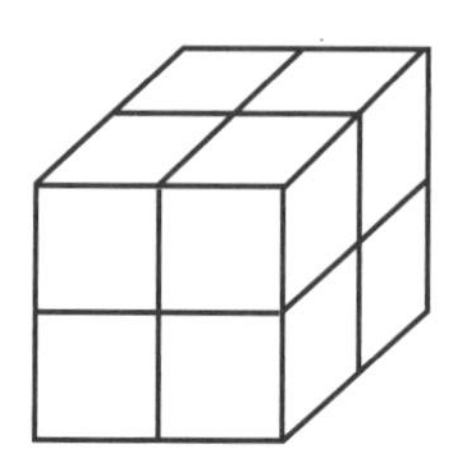

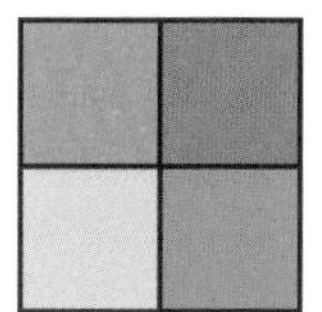 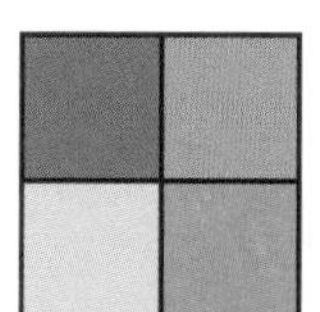 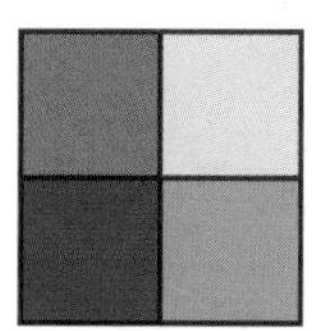 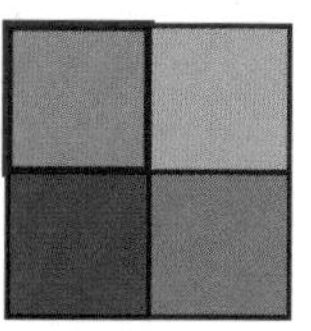 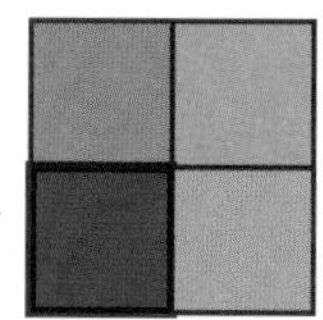 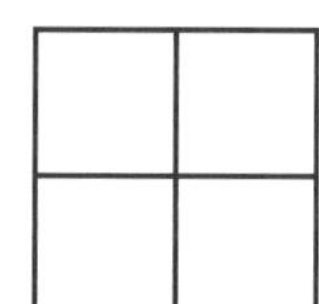

SOLUTION page 121

18 A 2x2x2 cube is made of eight differently coloured cubes of the same size. Four of the cubes have been removed. Five of the six faces are shown, as seen from above. A thicker border indicates that this cube is on the bottom row. When looking at the sixth and final face, as seen from above, what do you see?

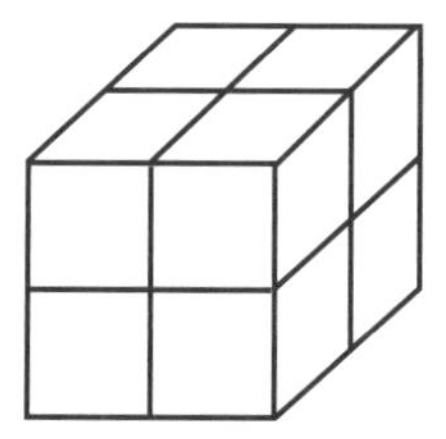

 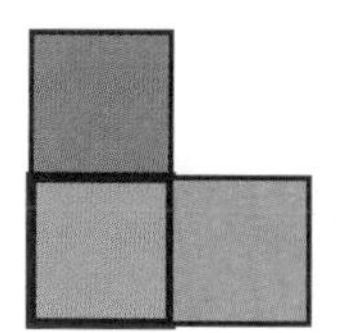 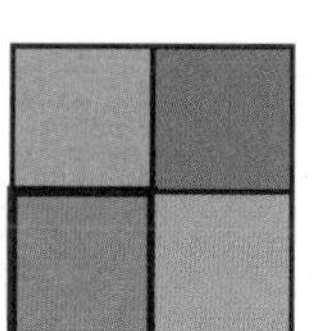 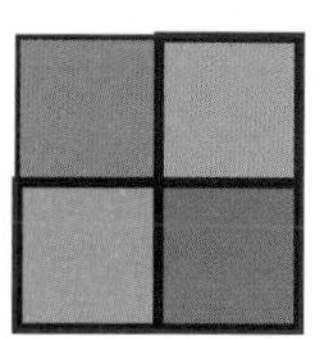 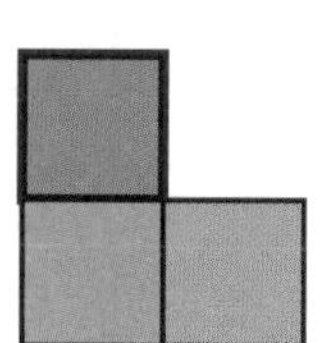 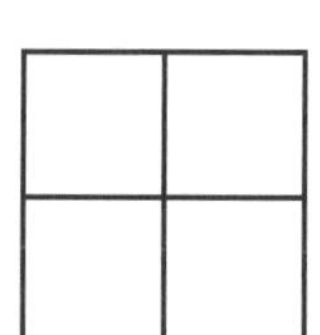

SOLUTION page 114

HOW MANY

How many discrete triangles can be seen in this star?

SOLUTION page 121

How many discrete triangles can be seen in this diagram?

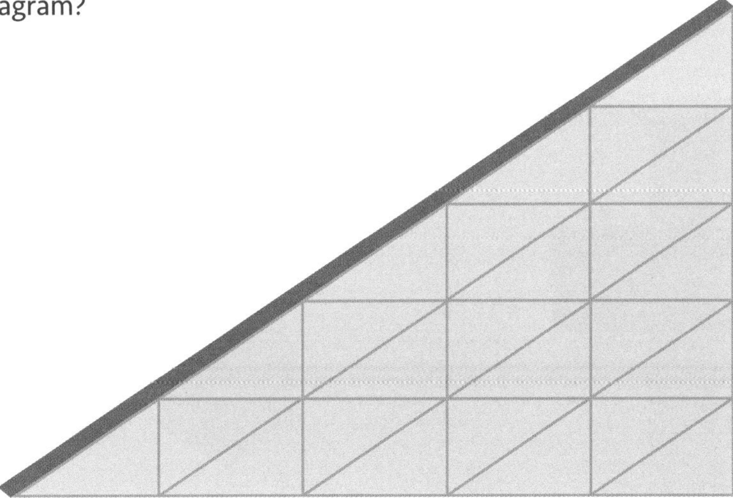

SOLUTION page 114

CUBE NET

21 How many, and which, of the five cubes shown can be formed from the cube net below?

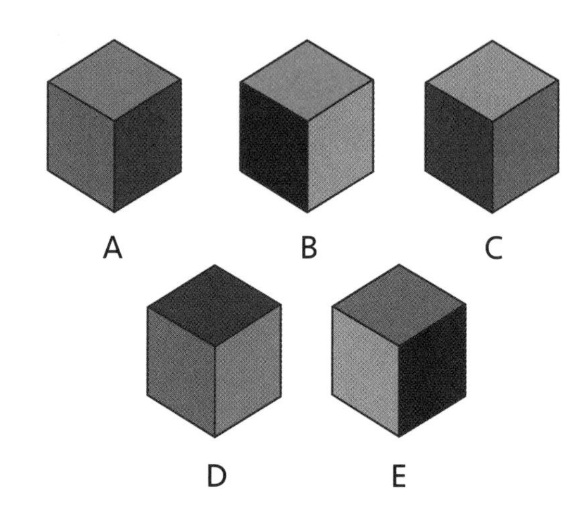

SOLUTION page 121

22 How many, and which, of the five cubes shown can be formed from the cube net below?

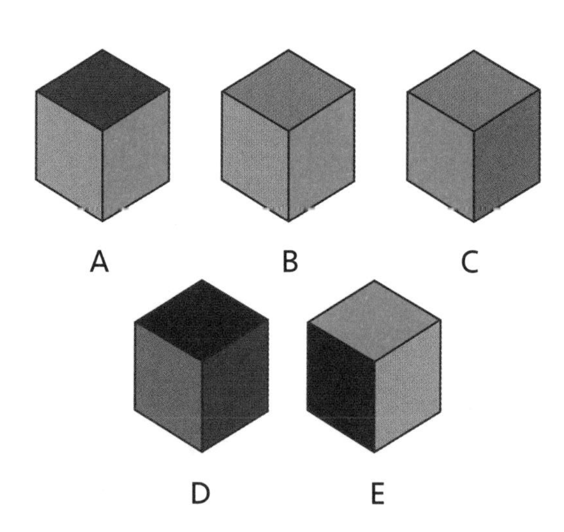

SOLUTION page 115

PERIMETER

Each brick in the wall has a perimeter measurement of 36cm. What is the overall perimeter of the wall?

SOLUTION page 114

The letter I is made of three identical blocks as shown. Using the same blocks to make the letter A, what is the overall perimeter of A, including the internal perimeter?

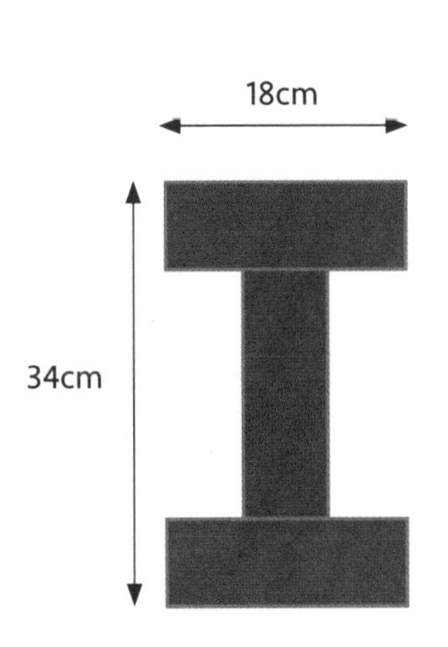

SOLUTION page 114

SHAPE UP

25 If a square sheet of paper is folded along the dotted lines, and cut along the solid lines, which pattern is produced? Cutting is done before folding.

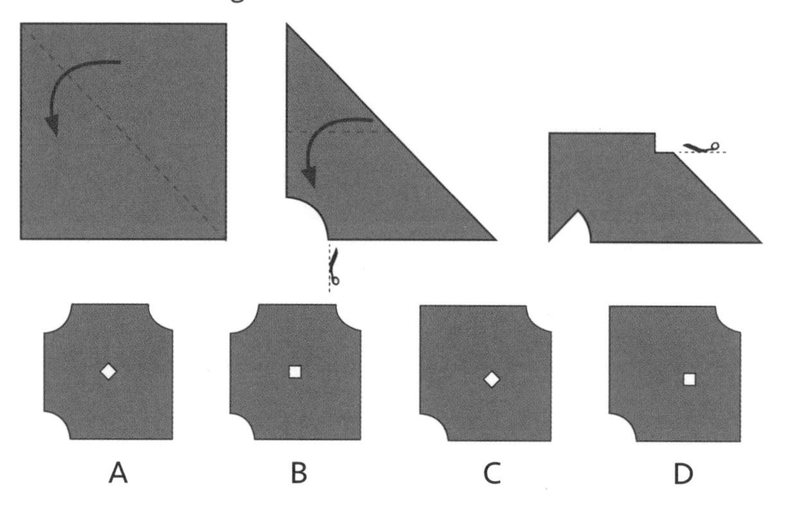

SOLUTION page 122

26 If a square sheet of paper is folded along the dotted lines, and cut along the solid lines, which pattern is produced? Cutting is done before folding.

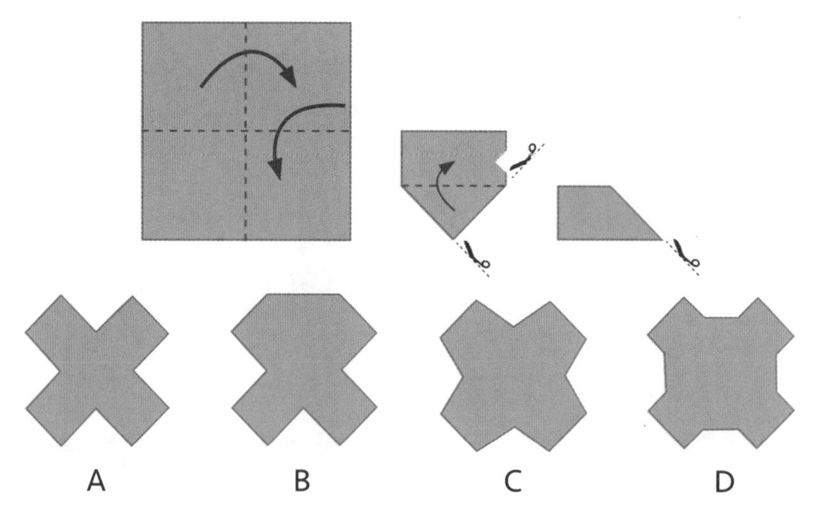

SOLUTION page 115

ANGLE GRINDER

27 From the information provided in the diagram, what is the value of the angle indicated by the question mark? Line AC intersects line BD and line EF at F.

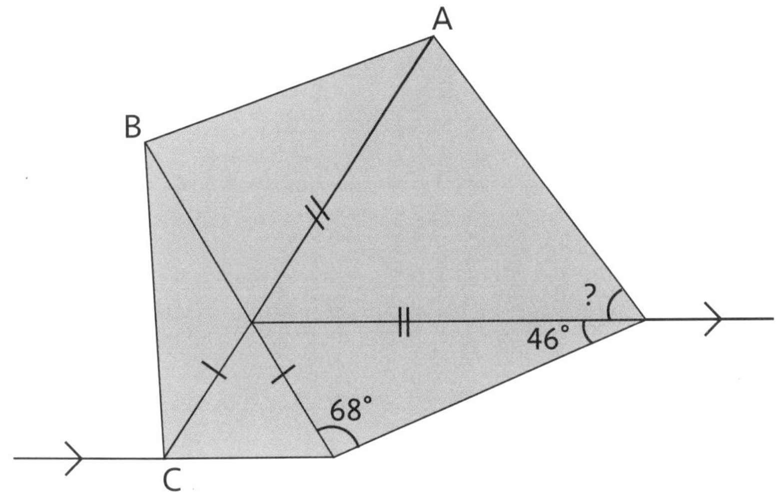

SOLUTION page 115

28 Line A bisects the regular pentagon pictured into two congruent halves. If line B is parallel to line A, what is the value of angle C?

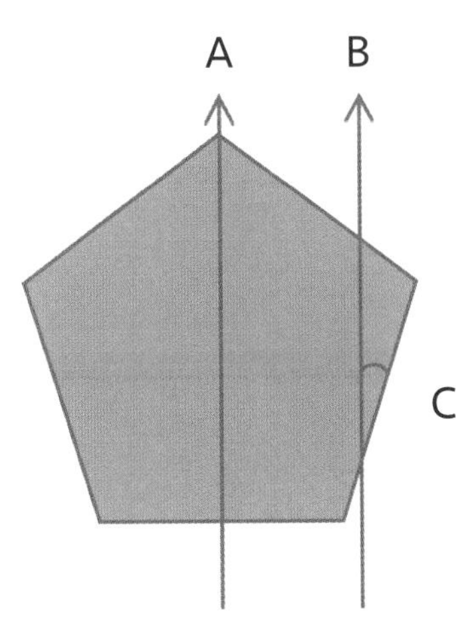

SOLUTION page 116

AREA MAZE (3D)

What area is represented by the question mark?

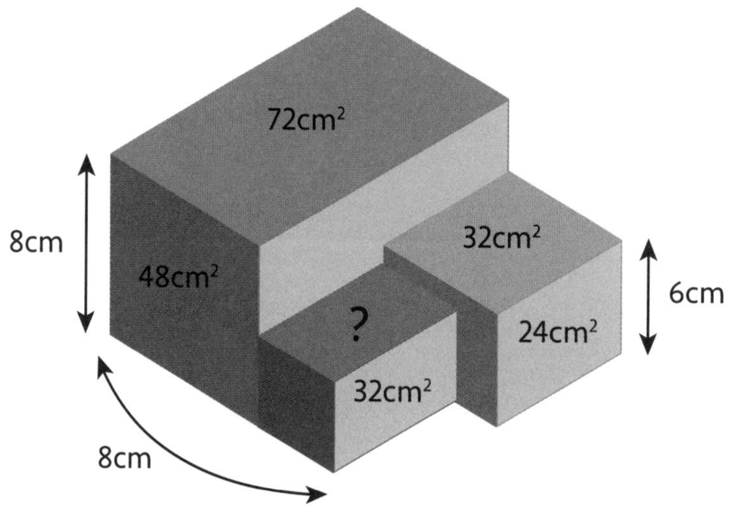

SOLUTION page 116

What area is represented by the question mark?

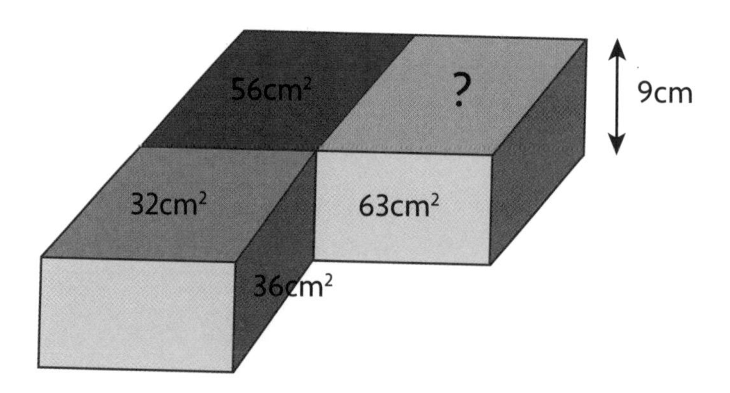

SOLUTION page 123

PENTOMINOES

31 Arrange six of the twelve pentominoes pictured (as they are and/or rotated and/or reflected) to create the pictured bridge.

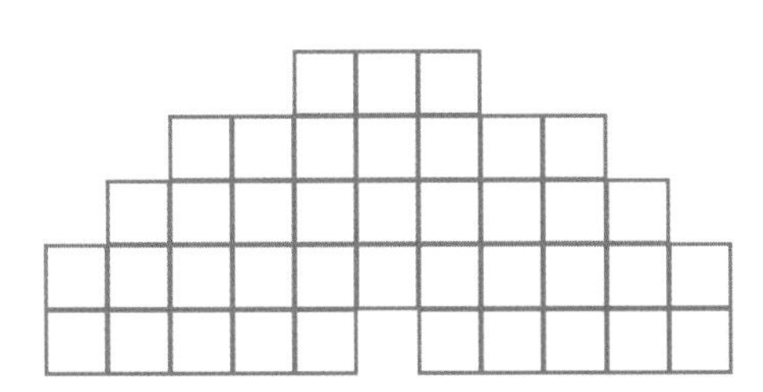

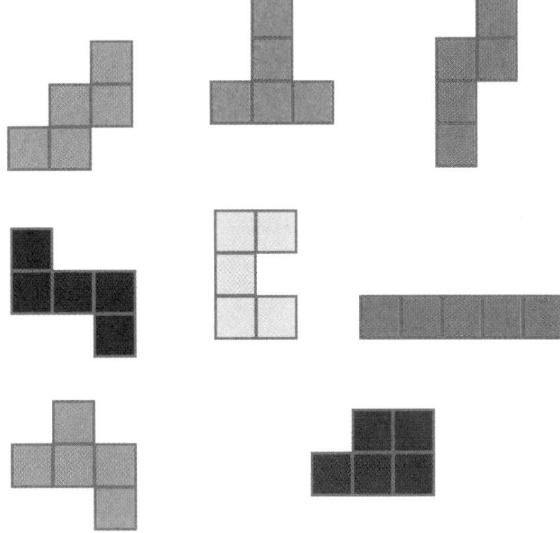

SOLUTION page 123

32 Arrange six of the twelve pentominoes pictured (as they are and/or rotated and/or reflected) to create the pictured key.

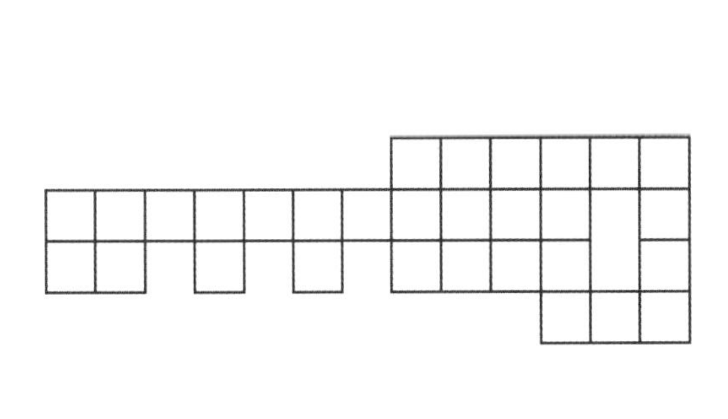

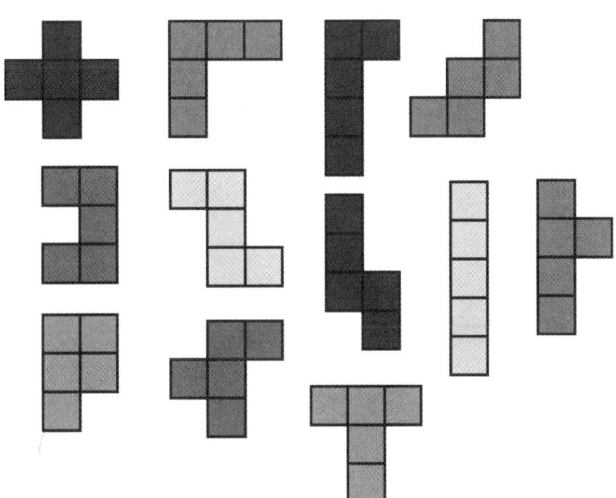

SOLUTION page 116

ARRANGEMENTS

Easy Money

A mathematician and her husband are on honeymoon in London. Walking along the South Bank, they come across a street entertainer with an irresistible offer: arrange nine £50 notes in ten rows of three notes. If you manage to do it, you get to keep the nine notes. The mathematician comes away £450 richer. How does she arrange the notes?

SOLUTION page 124

Blockhead

Alan the carpenter had prepared a rectangular piece of beech in order to make two identical kitchen tables for twin customers. He took a holiday, during which he received a call from his apprentice who explained that he'd had to cut out a smaller rectangle of unspecified size from the piece of beech in order to use for an urgent project. How can Alan cut the wood so that each twin gets an identical table?

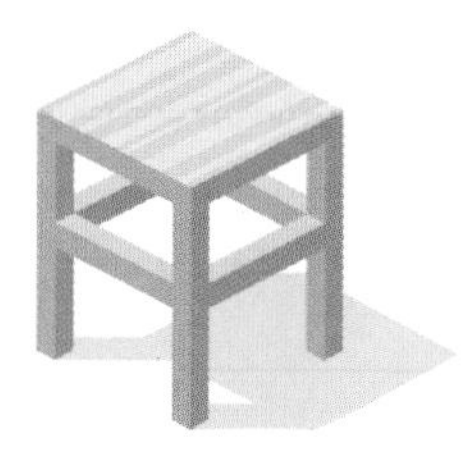

SOLUTION page 117

AREA MAZE

35 What is the length represented by the question mark?

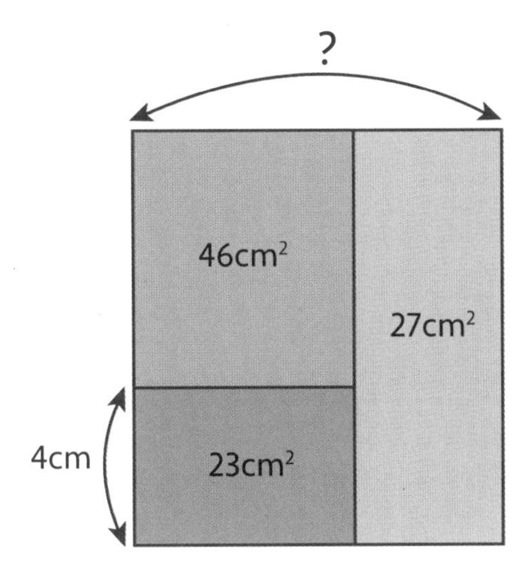

SOLUTION page 117

36 What is the length represented by the question mark?

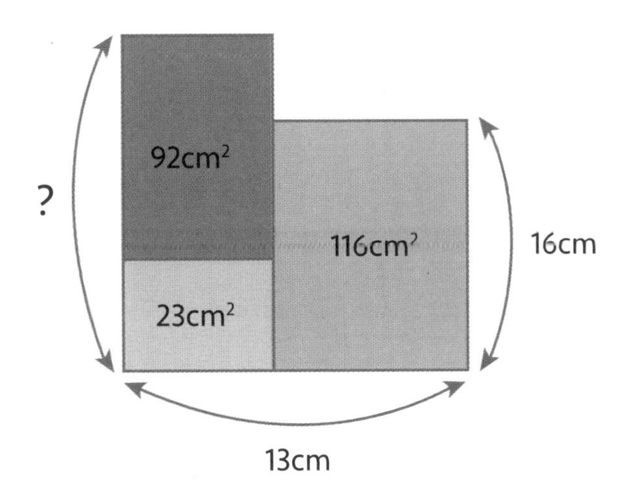

SOLUTION page 124

AREA MAZE (PERCENTAGES)

To the nearest percentage, what proportion of the octagon is represented by triangles?

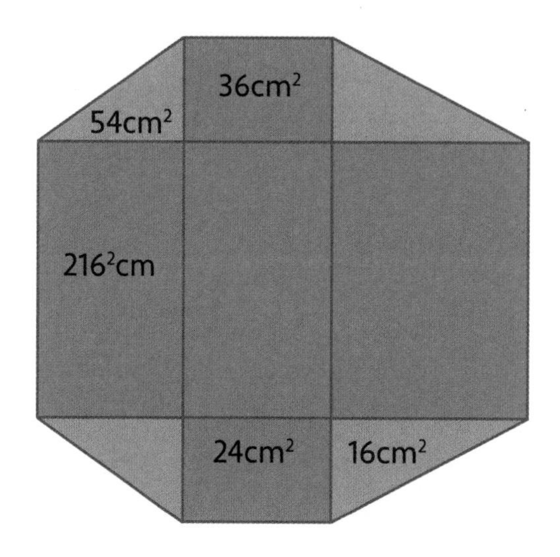

SOLUTION page 117

Which combined area is larger – blue or red?

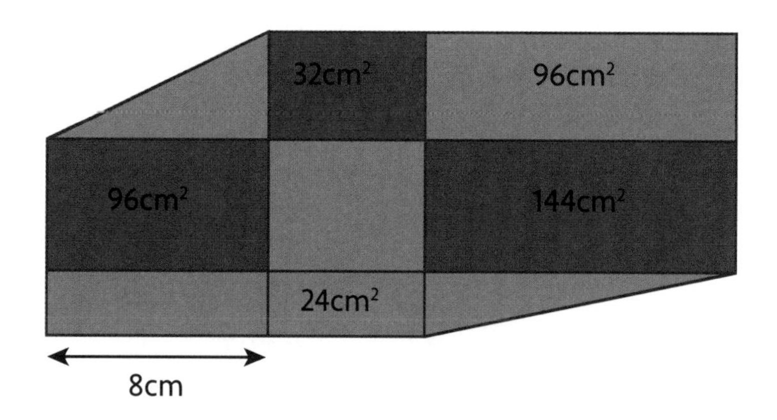

SOLUTION page 118

SPLIT DECISION

Split the grid into four identical shapes, following the marked lines. Shapes may be rotated, but not flipped.

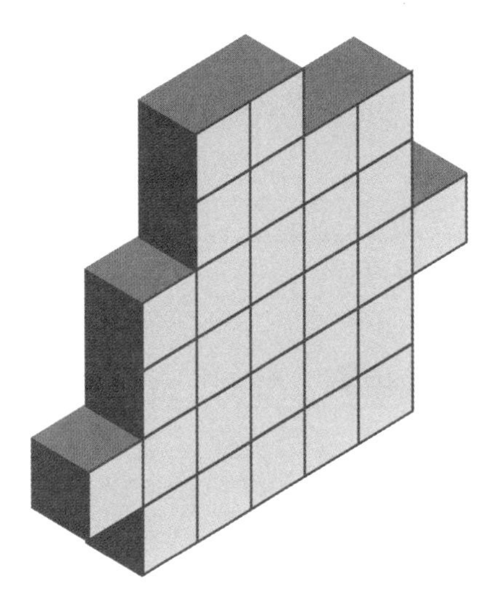

SOLUTION page 113

Split the grid into three identical shapes, following the marked lines. Shapes may be rotated, but not flipped.

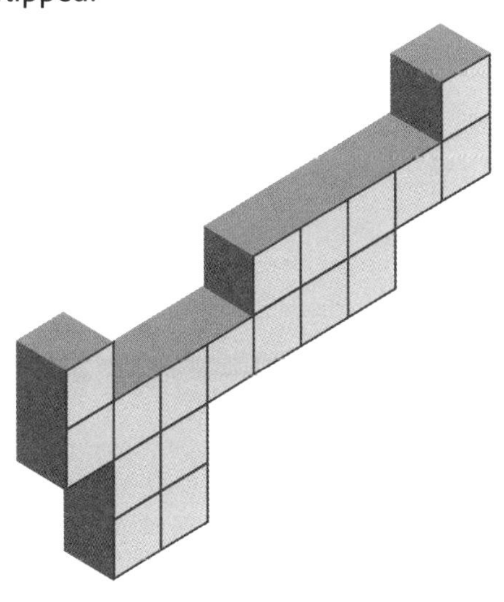

SOLUTION page 118

TANGRAM

41 The seven different shapes provided can be combined to form the outlined fox. How?

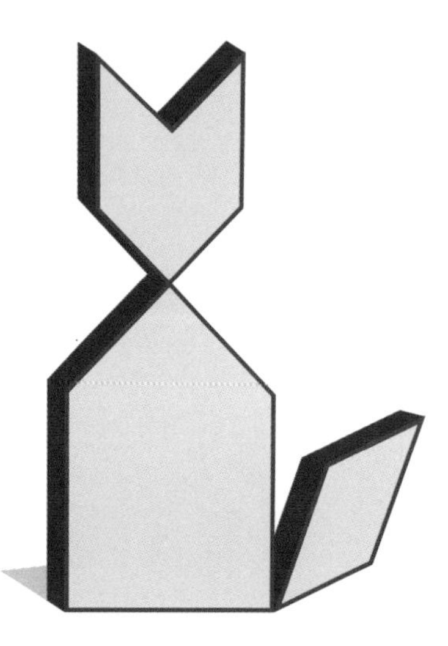

SOLUTION page 112

42 The seven different shapes provided can be combined to form the outlined rocket. How?

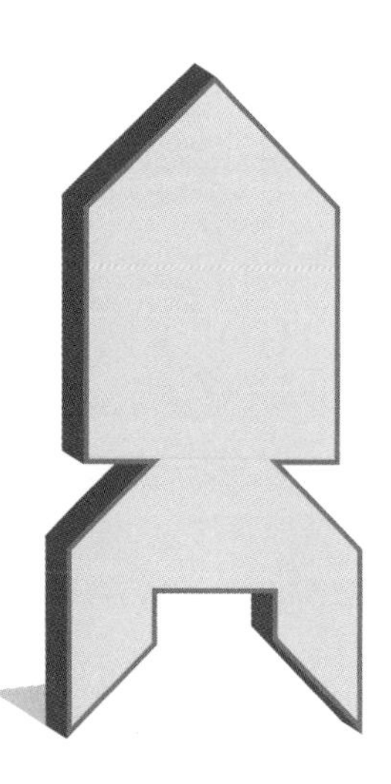

SOLUTION page 118

SPLIT DECISION

Split the grid into four identical shapes, following the marked lines. Shapes may be rotated, but not flipped.

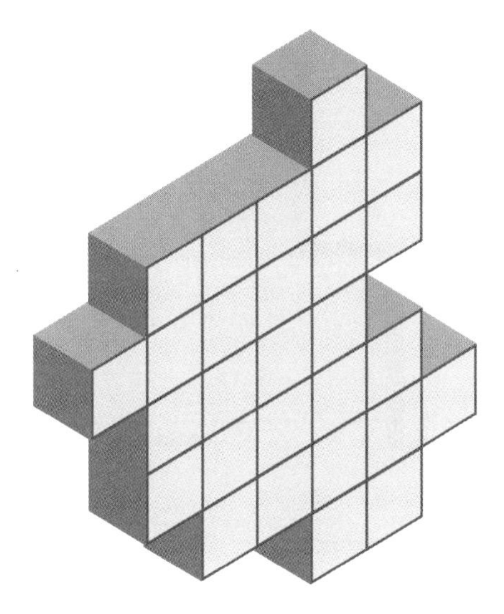

SOLUTION page 130

Split the grid into four identical shapes, following the marked lines. Shapes may be rotated, but not flipped.

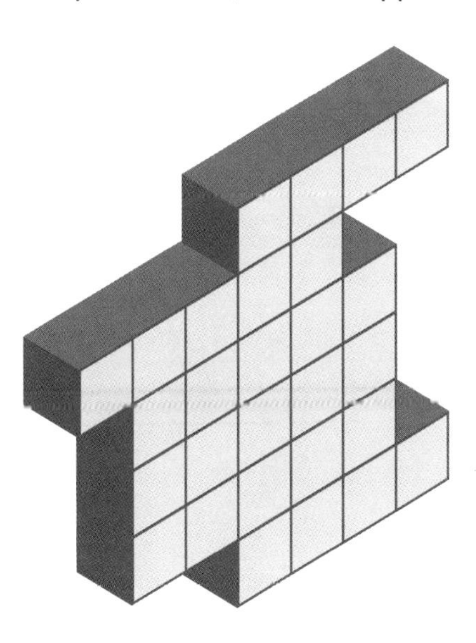

SOLUTION page 119

SQUARE DEAL

45 Divide this shape into two equal halves, marking the point of rotation.

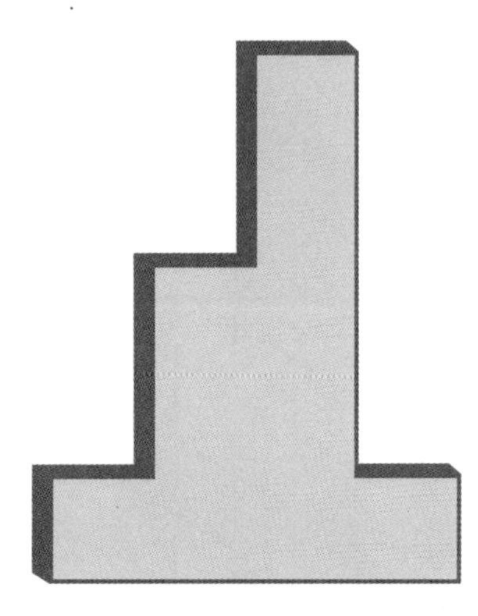

SOLUTION page 112

46 Split the shape into two pieces that can be arranged to form an 8x8 square.

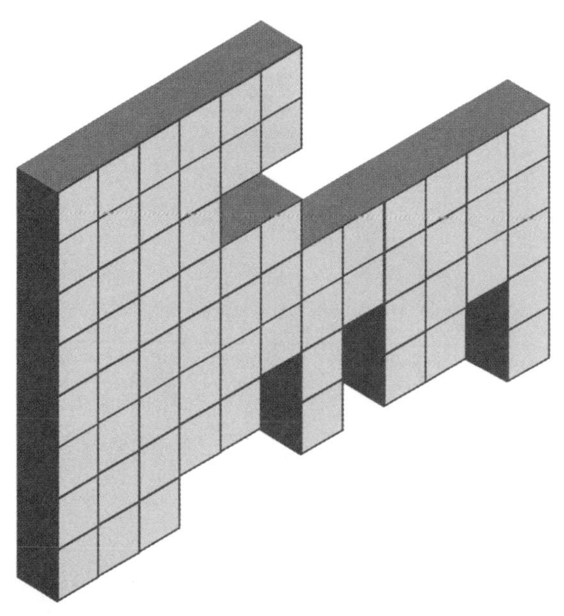

SOLUTION page 119

MATCHSTICKS

A) Move four matches to make three squares. Each match must form part of a square, and no matches can overlap.
B) Move three matches to leave three squares. Each match must form part of a square, and no matches can overlap.

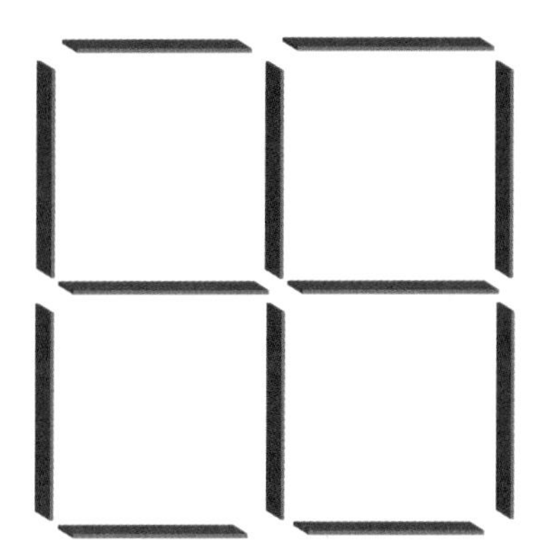

SOLUTION page 119

Move three matches to make three squares. Each match must form part of a square, and no matches can overlap.

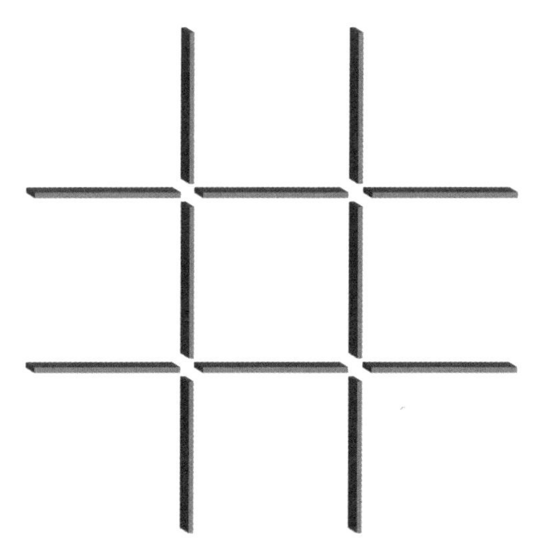

SOLUTION page 112

ADDING LINES

49 What is the maximum number of triangles that can be created by adding one triangle to this design? Where is it placed?

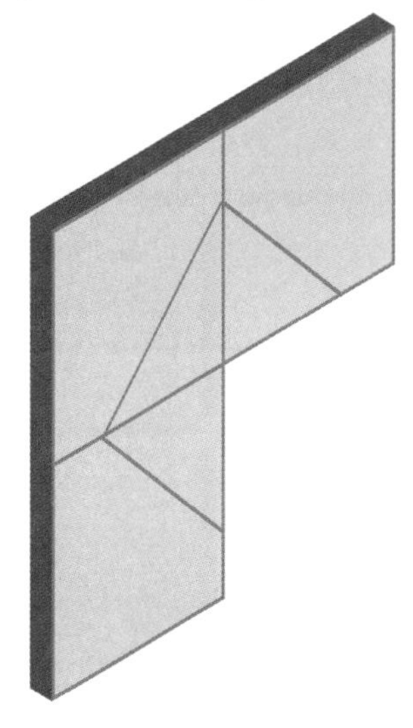

SOLUTION page 113

50 Add four straight lines, which don't touch, to create four triangles and eight squares.

SOLUTION page 120

LINE DRAWING

51 Use five lines, running from one side to another, to divide the rectangle into fourteen areas. Each area must contain one square.

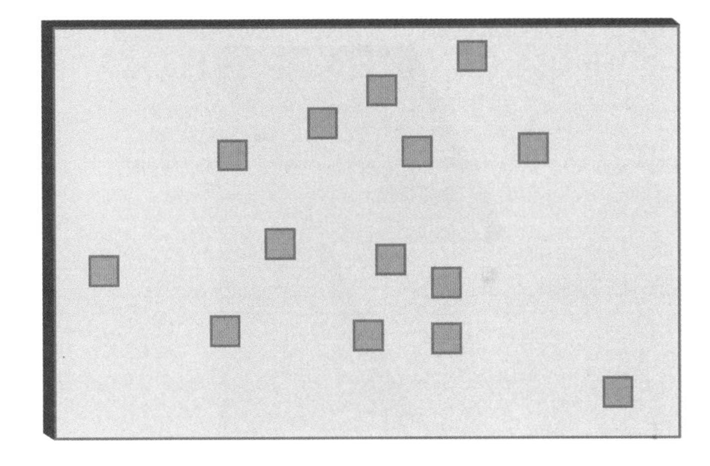

SOLUTION page 120

52 Use four lines, running from one side to another, to divide the rectangle into nine areas. Each area must contain one dot.

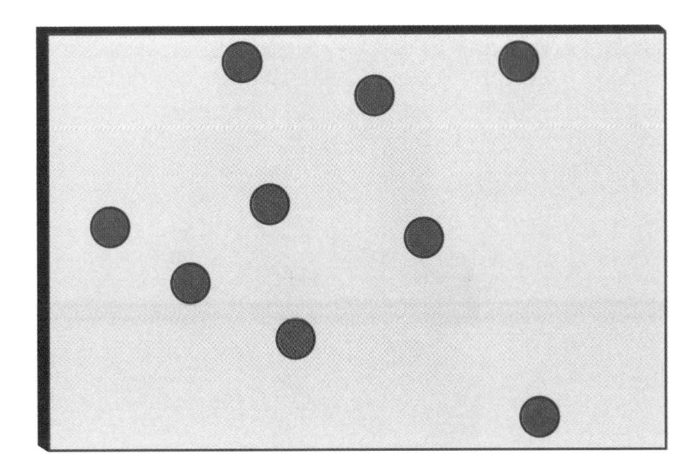

SOLUTION page 113

CONGRUENT SHAPES

53 Split the diagram into two congruent shapes, identical when rotated and/or flipped. There are no grid lines to follow, but there is enough information in the diagram.

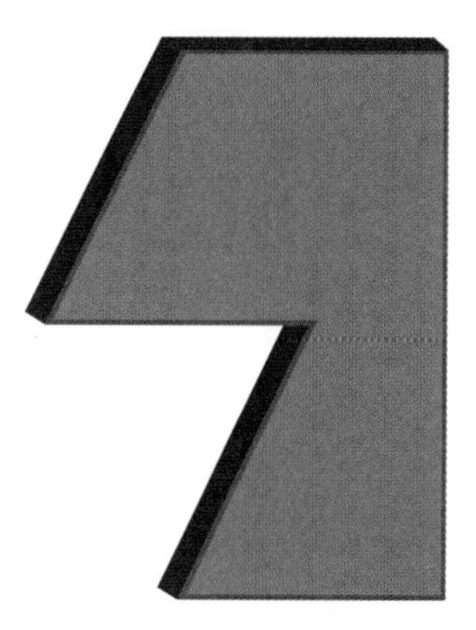

SOLUTION page 120

Split the grid into three congruent shapes, identical when rotated and/or flipped.

SOLUTION page 120

XOXO

55 Divide the grid into four congruent shapes, with the point of rotation at the centre of the grid. We have included four blue squares and three red squares as clues – no shape contains more than one blue square and one red square.

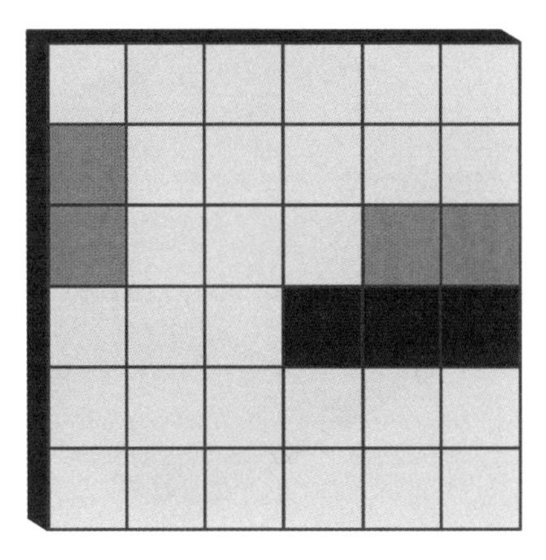

SOLUTION page 114

56 Divide the grid into four congruent shapes, with the point of rotation at the centre of the grid. We have included three blue squares as clues – no shape contains more than one blue square.

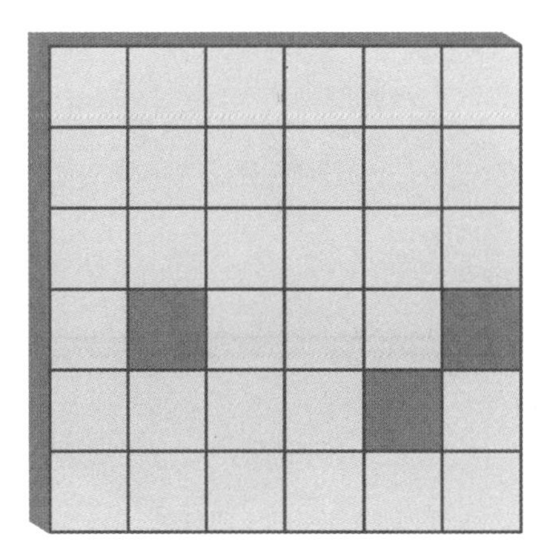

SOLUTION page 132

OVERLAID

57 What is the smallest number of overlaid paper squares that can be used to create this pattern?
In what order were they laid?

SOLUTION page 115

58 What is the smallest number of overlaid paper squares that can be used to create this pattern?
In what order were they laid?

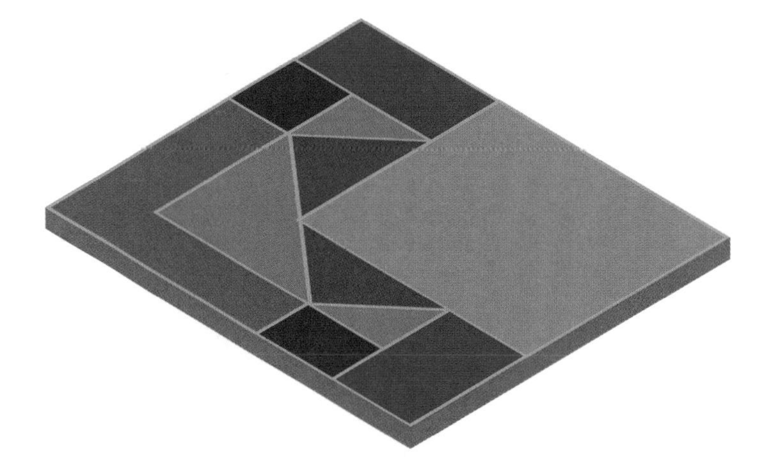

SOLUTION page 121

OVERVIEW

A 2x2x2 cube is made of eight differently coloured cubes of the same size. Three of the cubes have been removed. Five of the six faces are shown, as seen from above. A thicker border indicates that this cube is on the bottom row. When looking at the sixth and final face, as seen from above, what do you see?

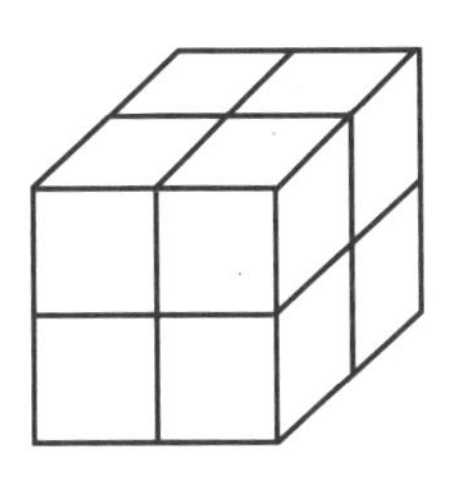

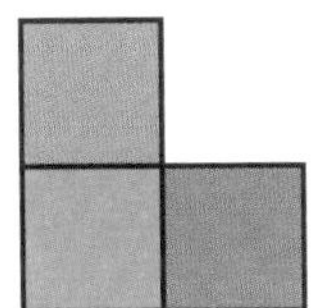
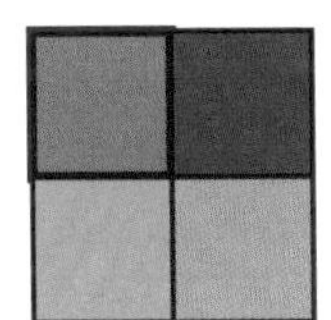
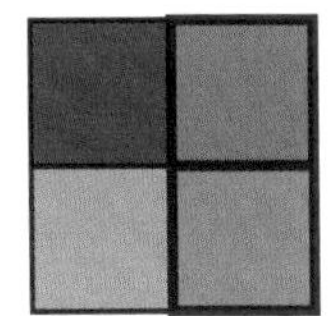
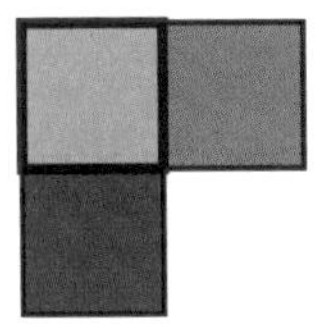

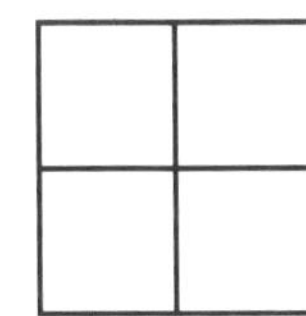

SOLUTION page 121

A 2x2x2 cube is made of eight differently coloured cubes of the same size. Two of the cubes have been removed. Five of the six faces are shown, as seen from above. A thicker border indicates that this cube is on the bottom row. When looking at the sixth and final face, as seen from above, what do you see?

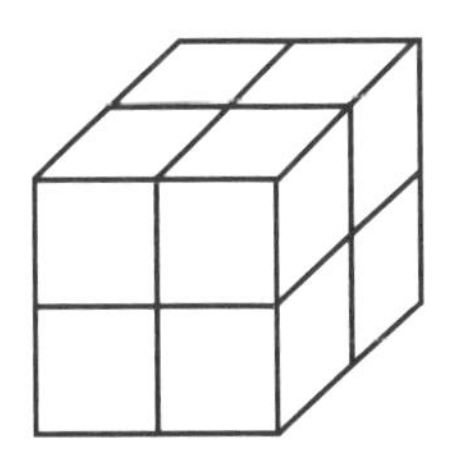

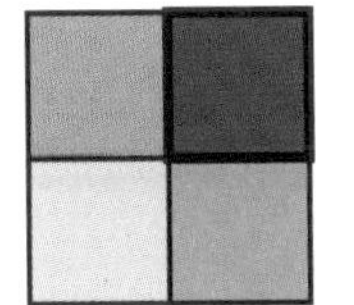
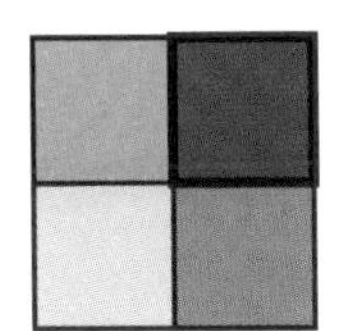
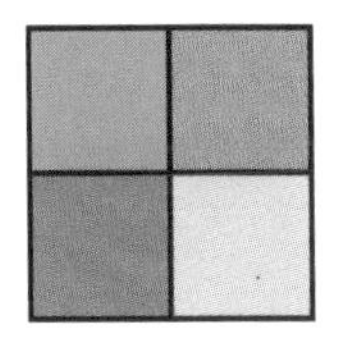
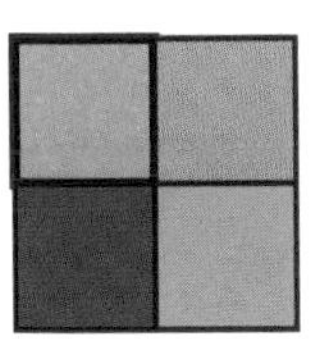
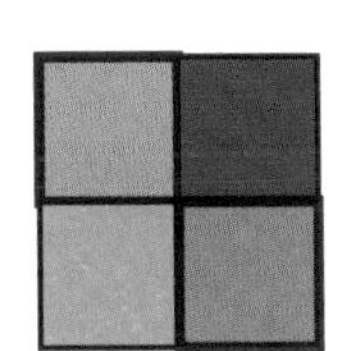
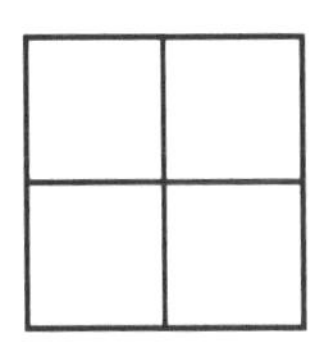

SOLUTION page 114

HOW MANY

61 How many discrete diamonds can be seen in this diagram?

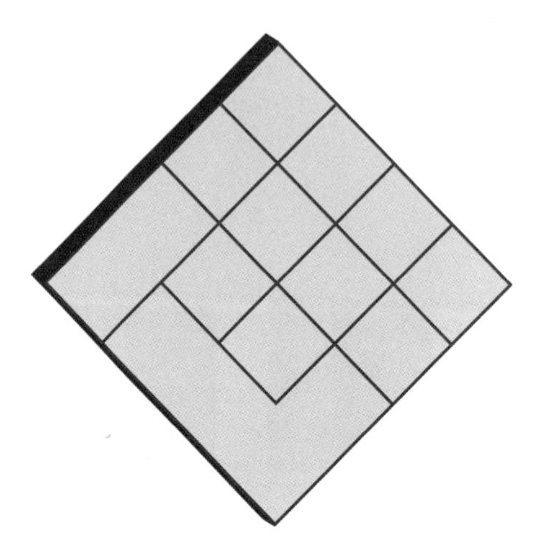

SOLUTION page 114

62 How many discrete triangles can be seen in this diagram?

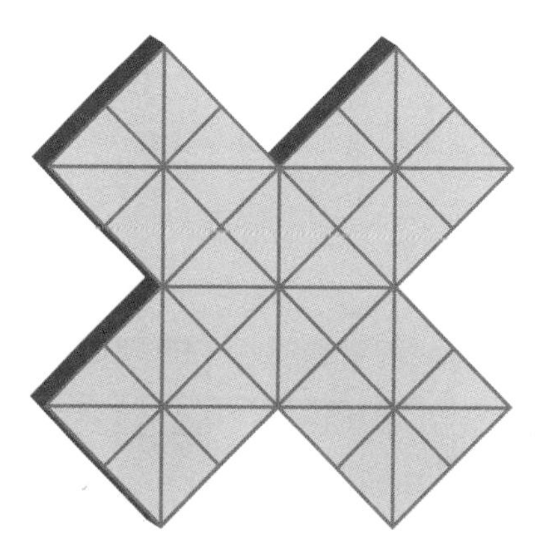

SOLUTION page 121

CUBE NET

63 How many, and which, of the five cubes shown can be formed from the cube net below?

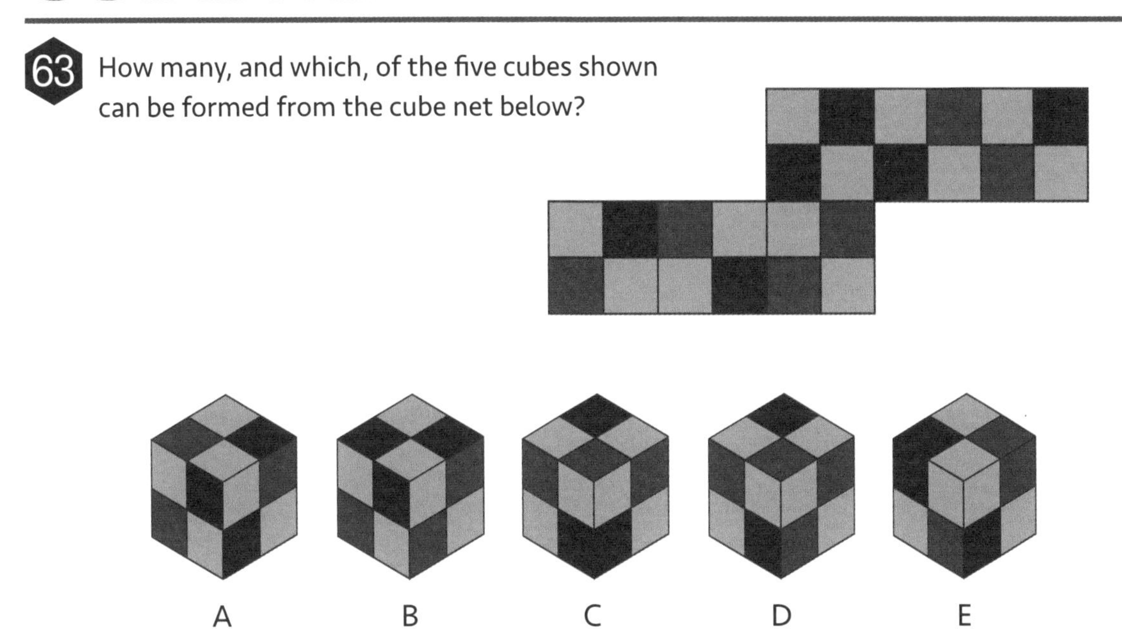

SOLUTION page 121

64 How many, and which, of the five cubes shown can be formed from the cube net below?

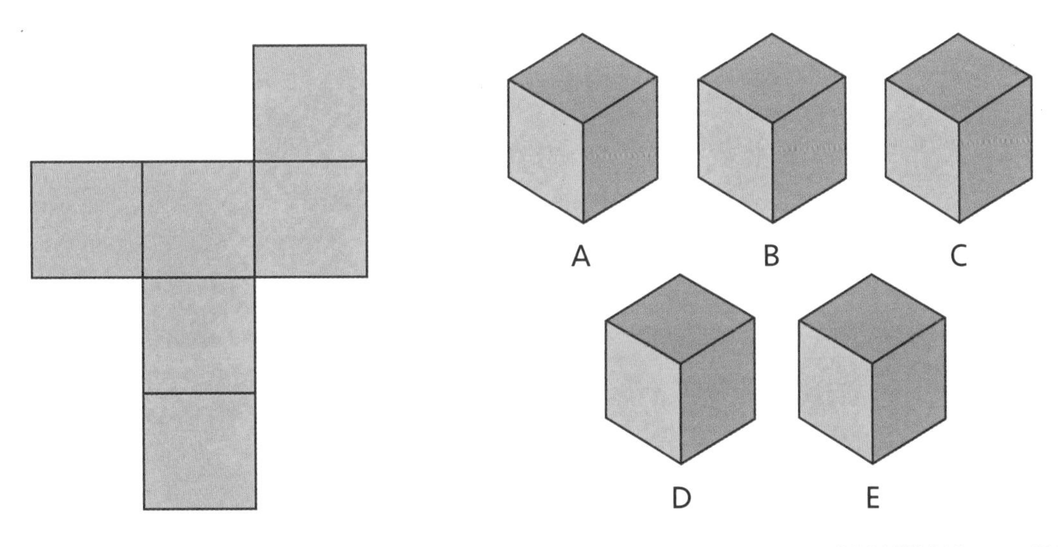

SOLUTION page 114

PERIMETER

65 A tiler uses three differently shaped tiles to make the pattern shown. The diamond tile has a perimeter of 28cm. What is the perimeter of the yellow-tiled area, and what percentage is it of the overall perimeter?

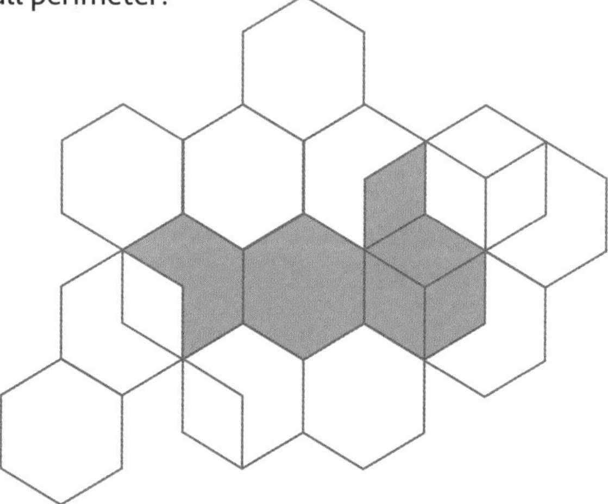

SOLUTION page 121

66 Each block in the parquet flooring is twice as long as wide. If the overall perimeter of this area is 504cm, how wide is each block?

SOLUTION page 122

SHAPE UP

If a square sheet of paper is folded along the dotted lines, and cut along the solid lines, which pattern is produced? Cutting is done before folding.

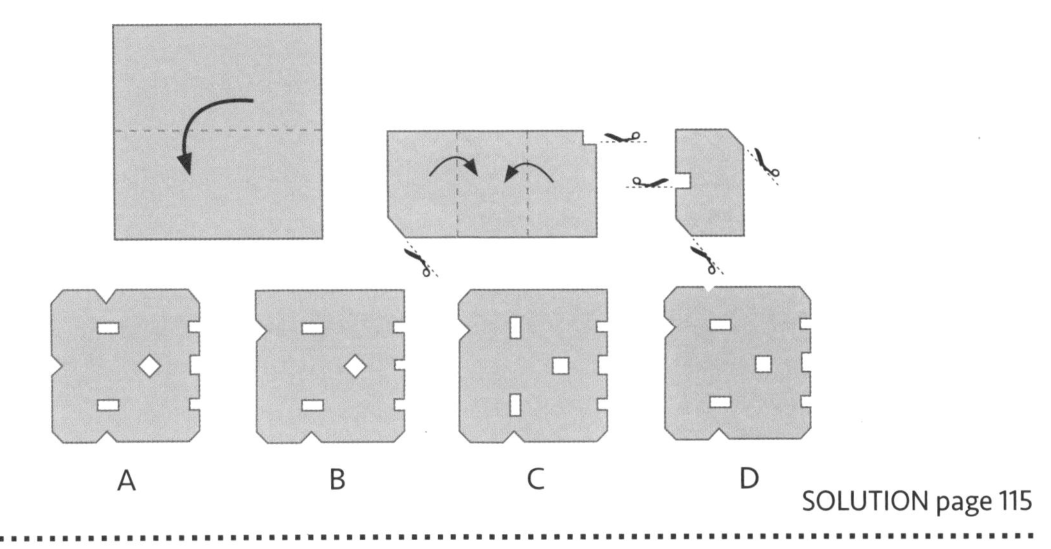

SOLUTION page 115

If a square sheet of paper is folded along the dotted lines, and cut along the solid lines, which pattern is produced? Cutting is done before folding.

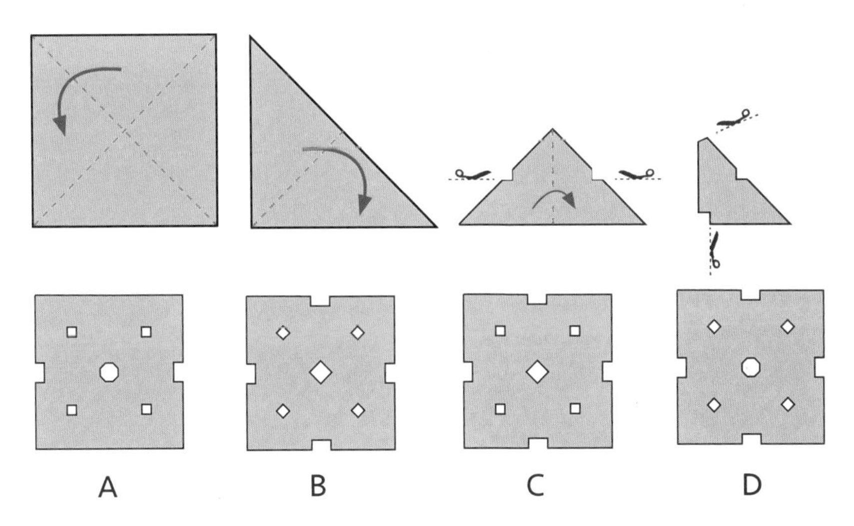

SOLUTION page 122

ANGLE GRINDER

69 If angle A and angle B are given as shown, how many degrees is angle C?

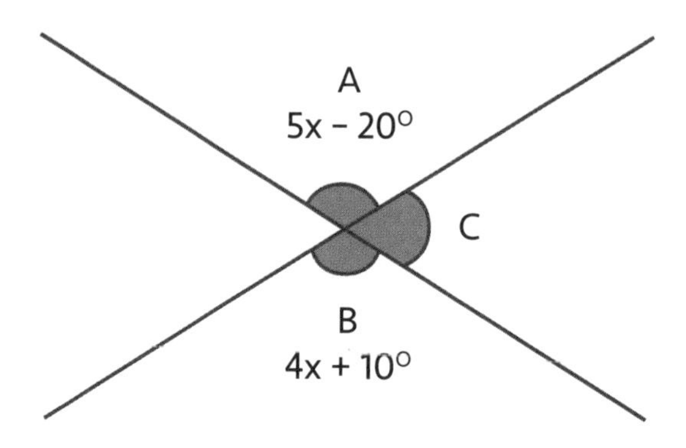

SOLUTION page 122

70 What is the value of the angle marked by the question mark?

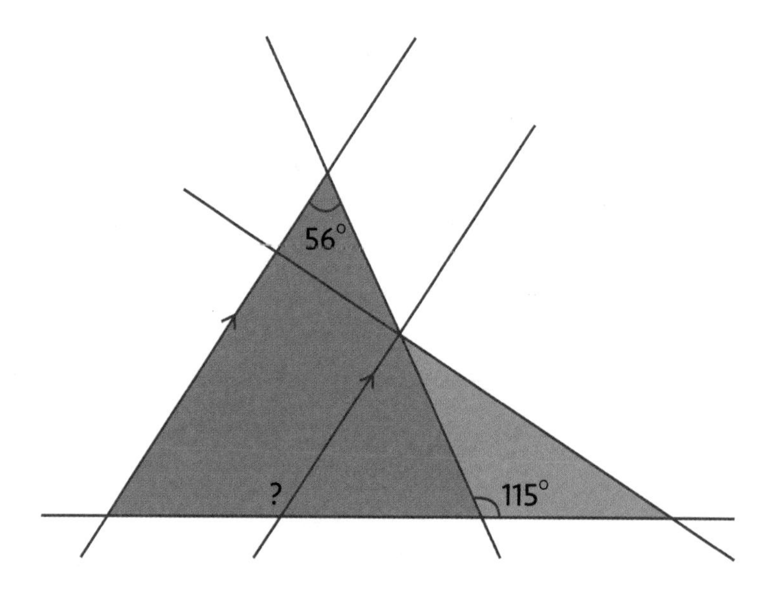

SOLUTION page 122

AREA MAZE (3D)

What length is represented by the question mark?

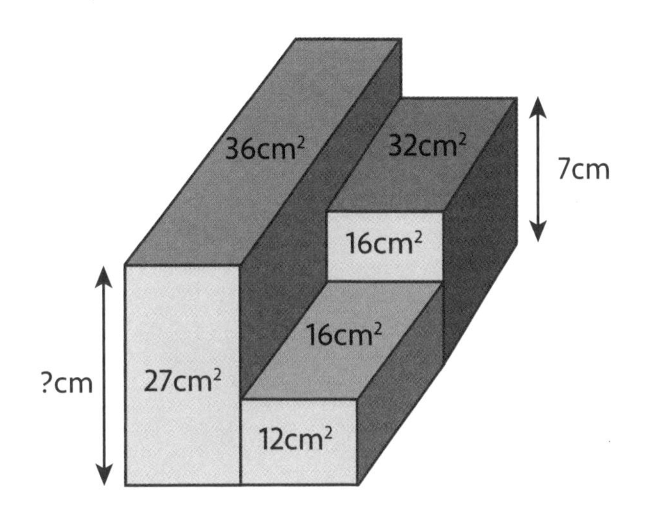

SOLUTION page 122

72 What length is represented by the question mark?

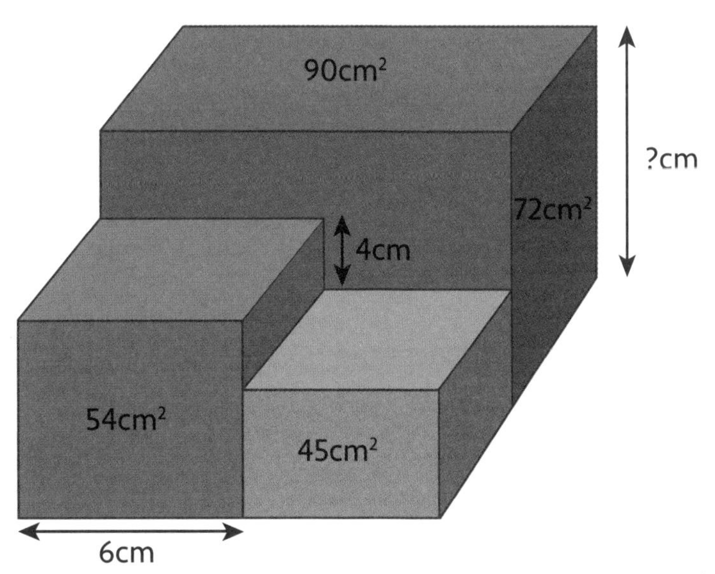

SOLUTION page 116

PENTOMINOES

73 Arrange seven of the twelve pentominoes pictured (as they are and/or rotated and/or reflected) to create the pictured musical note.

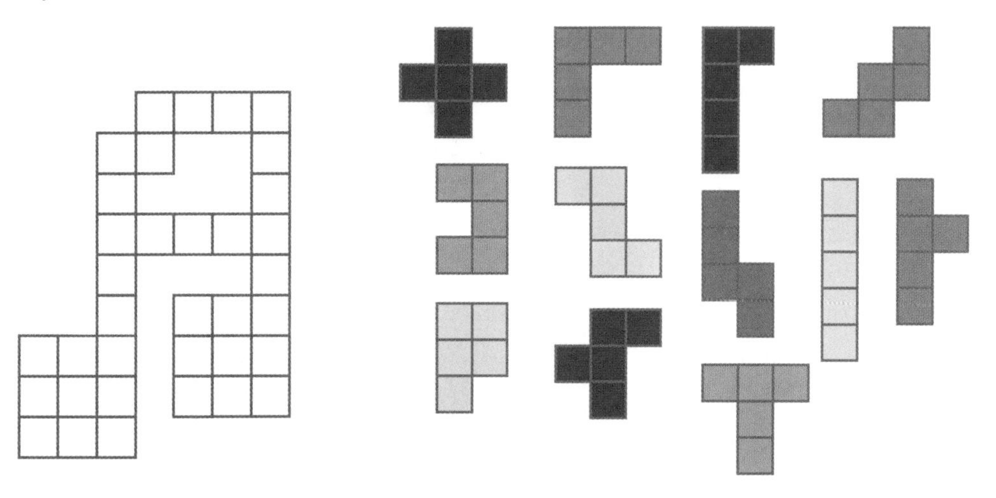

SOLUTION page 116

74 Arrange eight of the twelve pentominoes pictured (as they are and/or rotated and/or reflected) to create the pictured Great British Pound symbol.

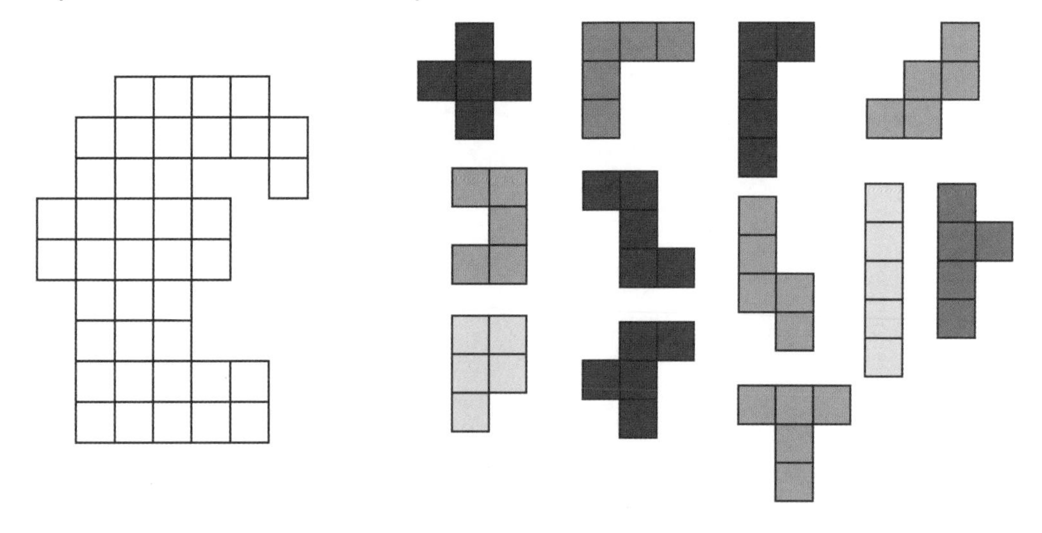

SOLUTION page 123

AREA MAZE (PERCENTAGES)

75 What proportion of the whole rectangle is represented by the grey rectangle?

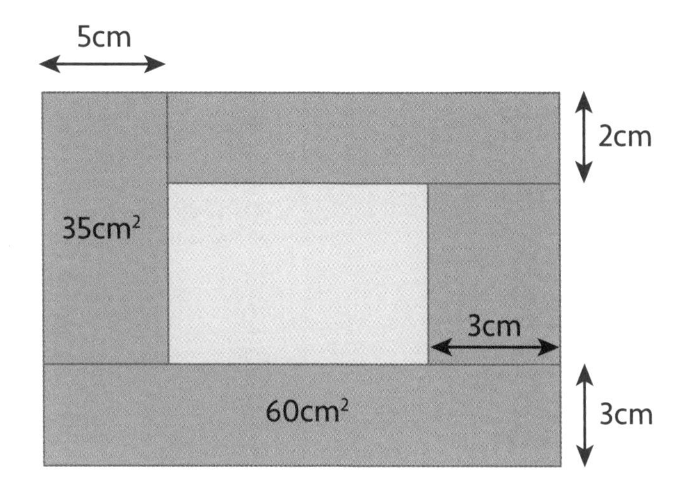

SOLUTION page 123

76 Which combined area is larger – white or orange?

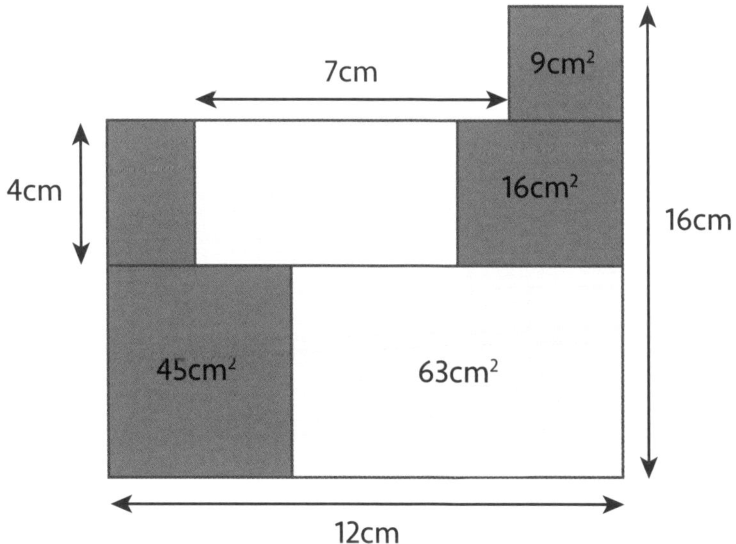

SOLUTION page 130

AREA MAZE

77 What is the area represented by the question mark?

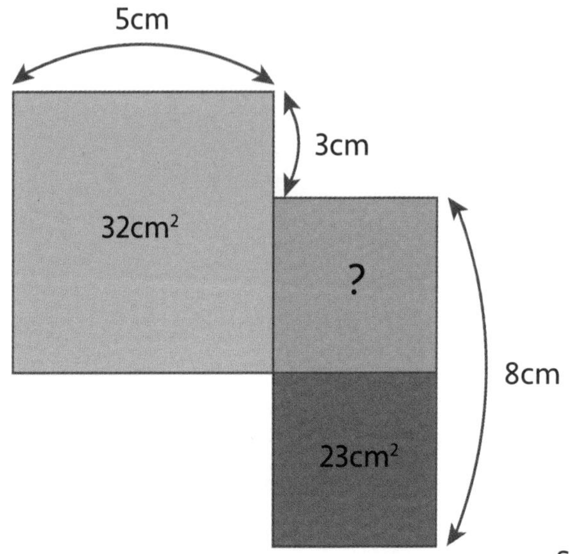

SOLUTION page 124

78 What is the length represented by the question mark?

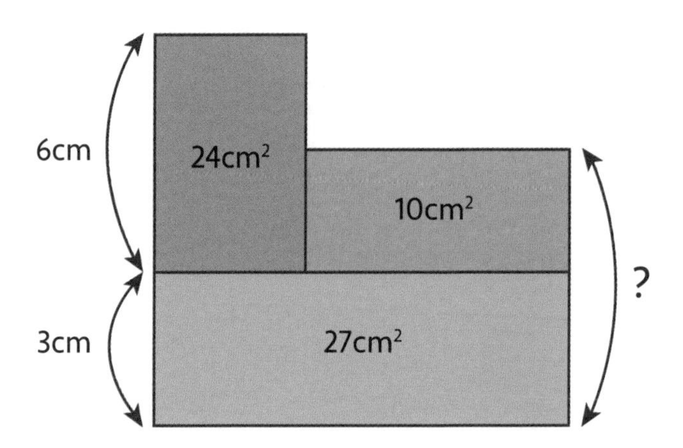

SOLUTION page 117

ARRANGEMENTS

Rock Hard

Curating an exhibition on the solar system, you have ten meteorites to display on plinths. You have a brainwave, and explain the arrangement of plinths simply by saying that the ten meteorites must be arranged in five rows of four. How is this achieved, and why?

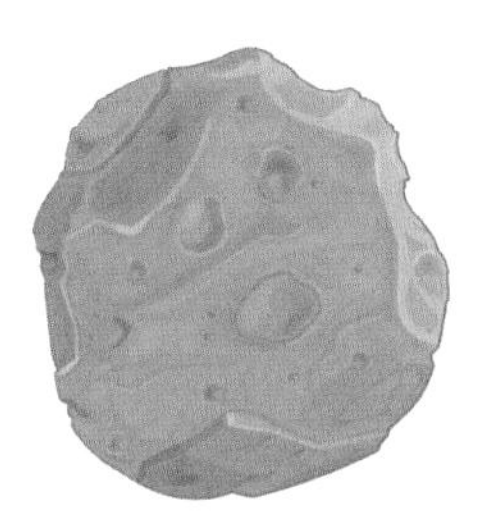

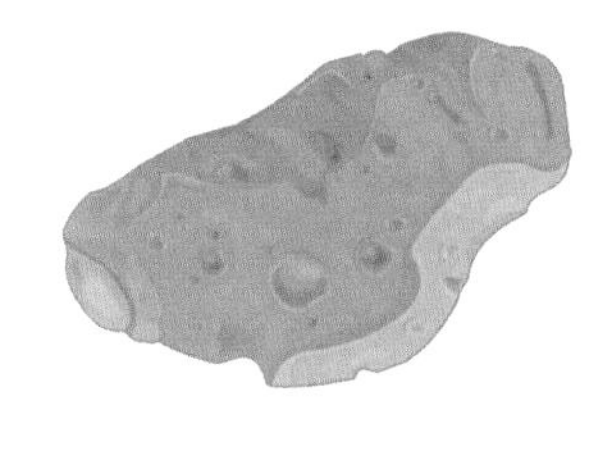

SOLUTION page 117

Building Blocks

Your architecture firm has won a bid to design seven new blocks of flats. All of the other bidders couldn't resolve one of the conditions of the brief – between them, the seven blocks of flats must form six rows of three blocks. How did you manage to arrange the blocks of flats?

SOLUTION page 124

TANGRAM

The seven different shapes provided can be combined to form the outlined boat. How?

SOLUTION page 130

The seven different shapes provided can be combined to form the outlined house. How?

SOLUTION page 125

SQUARE DEAL

83 Divide this shape into two equal halves, marking the point of rotation.

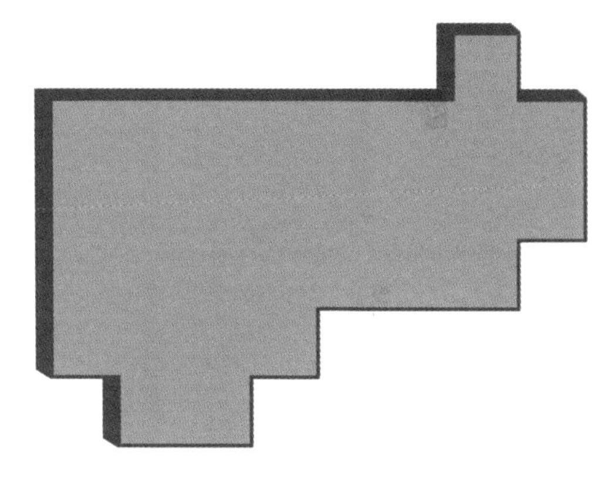

SOLUTION page 127

84 Split the shape into two pieces that can be arranged to form an 8x8 square.

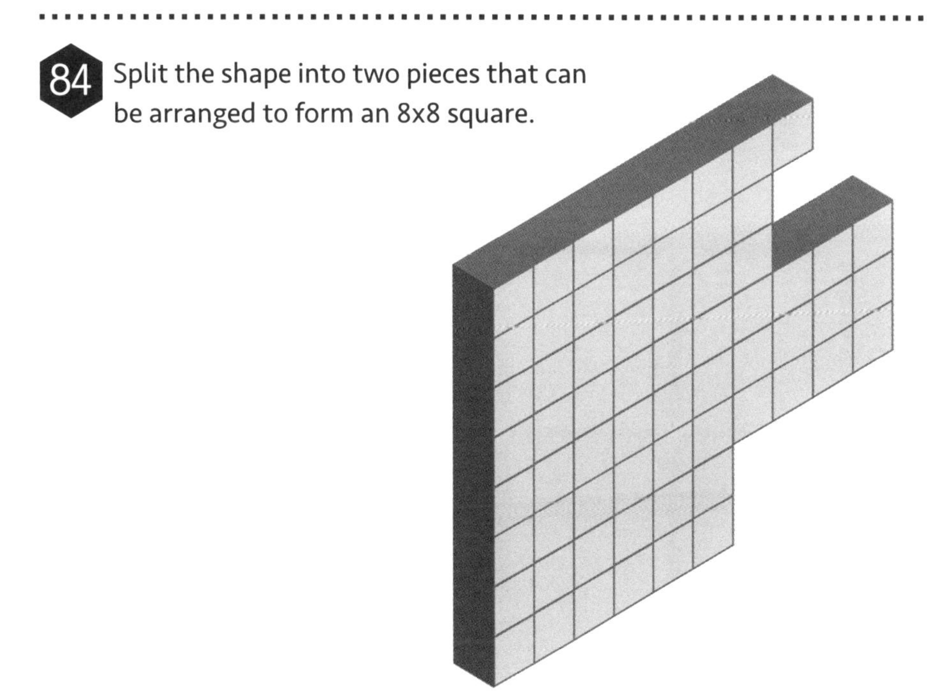

SOLUTION page 125

MATCHSTICKS

A) Remove two matches to make two squares.
B) Move two matches to make seven squares.

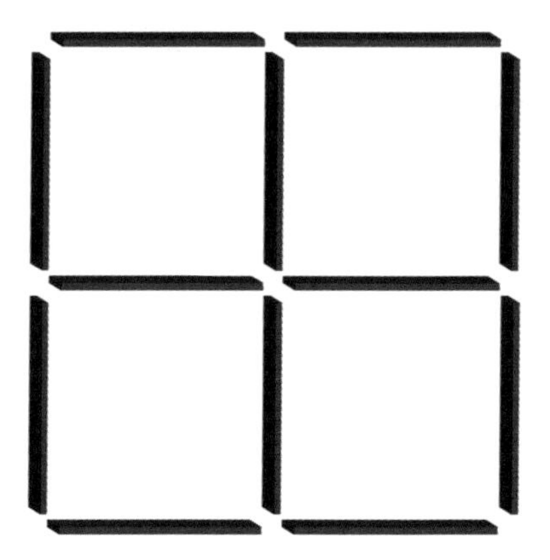

SOLUTION page 125

Move three matches to make four squares. Each match must form part of a square, and no matches can overlap.

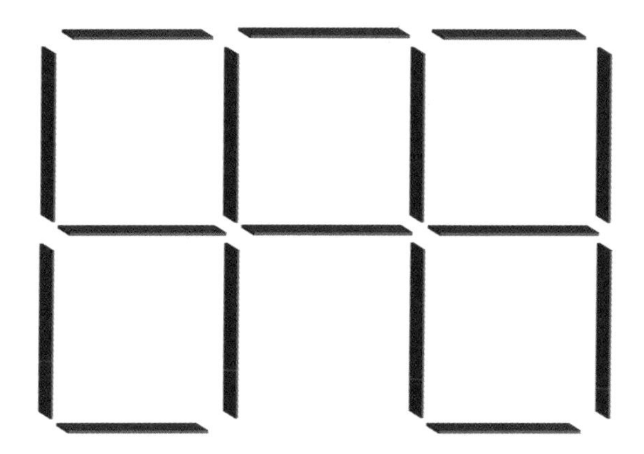

SOLUTION page 137

ADDING LINES

What is the maximum number of triangles that can be created by adding two squares to this design? How are the squares arranged?

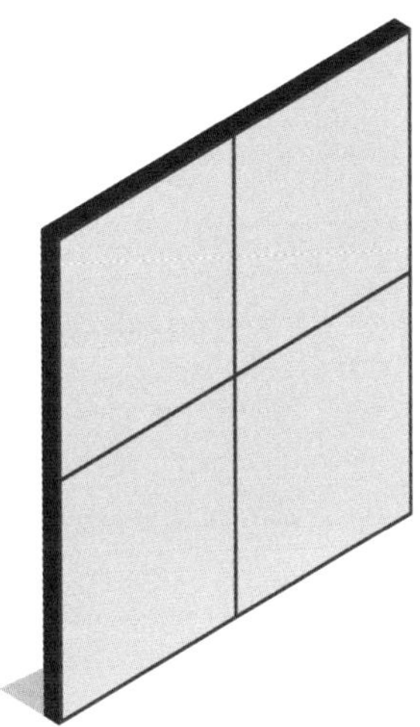

SOLUTION page 125

What regular polygon can be added to this pattern to create a chain of five-pointed stars?

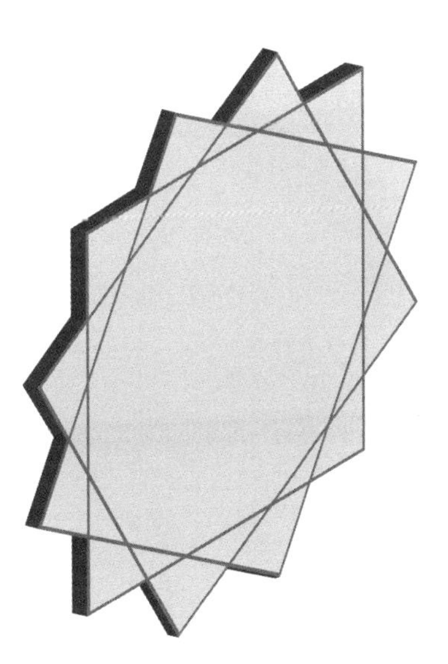

SOLUTION page 125

CONGRUENT SHAPES

Split the grid into four congruent shapes, identical when rotated and/or flipped.

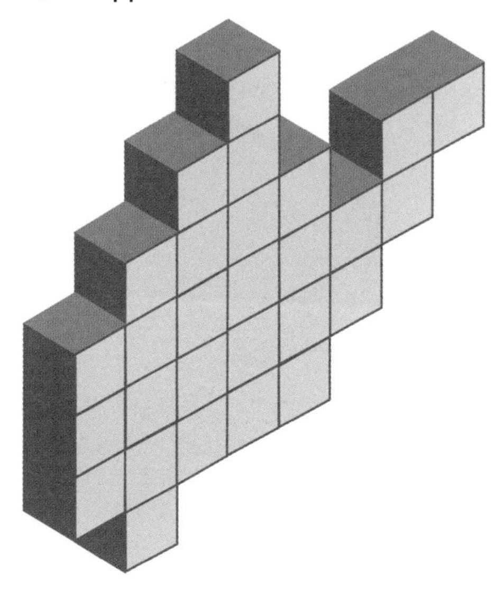

SOLUTION page 126

Split the grid into four congruent shapes, identical when rotated and/or flipped.

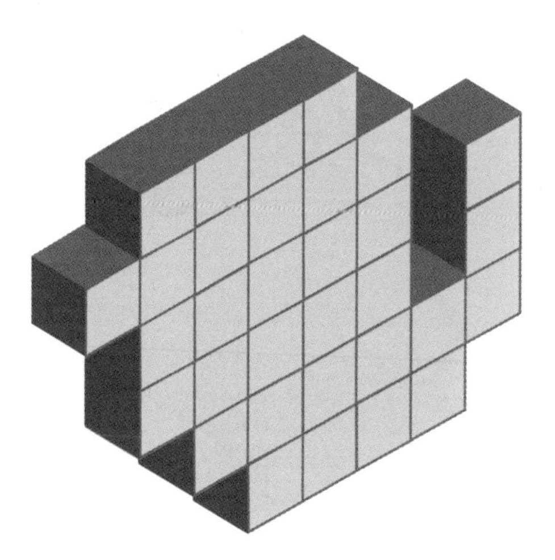

SOLUTION page 129

XOXO

91 Divide the grid into four congruent shapes, with the point of rotation at the centre of the grid. We have included blue, red and yellow squares as clues – no shape contains more than one of each.

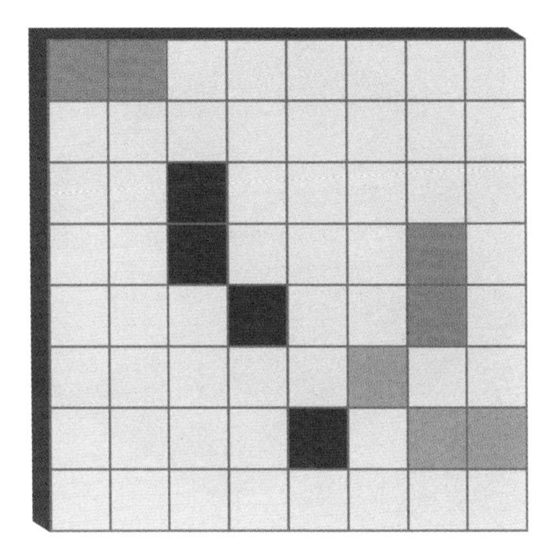

SOLUTION page 120

Divide the grid into four congruent shapes, with the point of rotation at the centre of the grid. We have included blue, red and yellow squares as clues – no shape contains more than one of each.

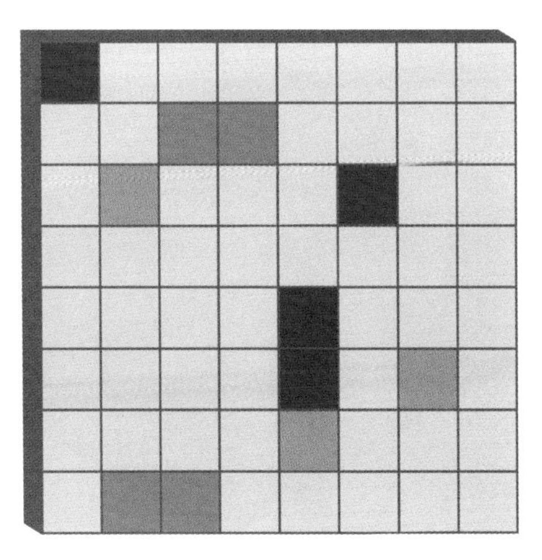

SOLUTION page 126

OVERLAID

93 What is the smallest number of overlaid paper squares that can be used to create this pattern? In what order were they laid?

SOLUTION page 126

94 What is the smallest number of overlaid paper squares that can be used to create this pattern? In what order were they laid?

SOLUTION page 121

OVERVIEW

A 2x2x2 cube is made of eight differently coloured cubes of the same size. Three of the cubes have been removed. Five of the six faces are depicted, as seen from above. A thicker border indicates that this cube is on the bottom row. When looking at the sixth and final face, as seen from above, what do you see?

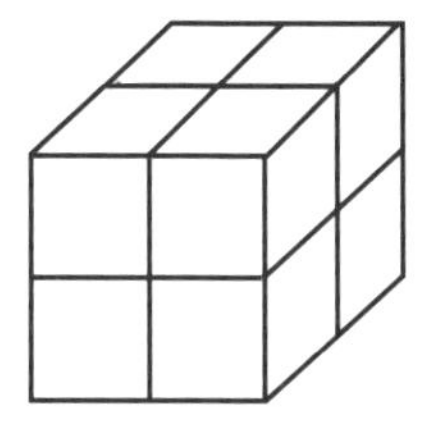

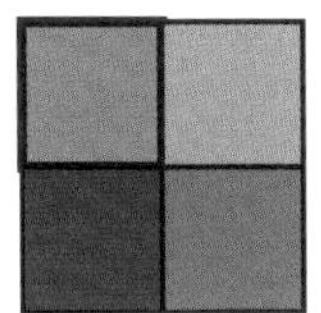 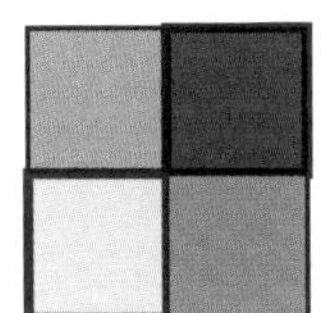 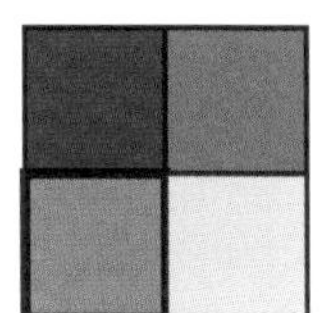 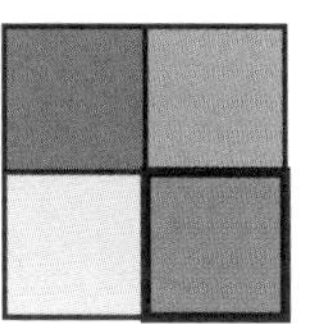 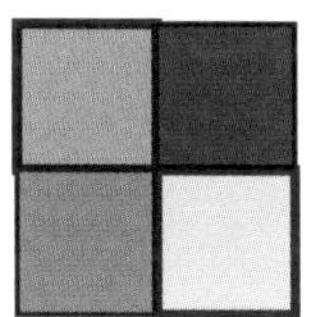 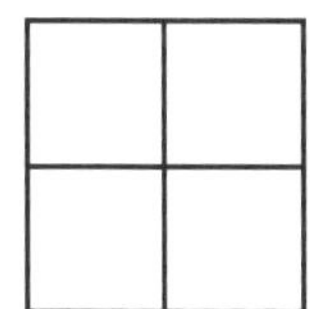

SOLUTION page 126

A 2x2x2 cube is made of eight differently coloured cubes of the same size. Three of the cubes have been removed. Five of the six faces are shown, as seen from above. A thicker border indicates that this cube is on the bottom row. When looking at the sixth and final face, as seen from above, what do you see?

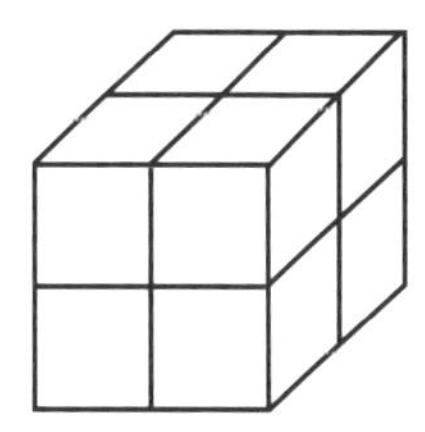

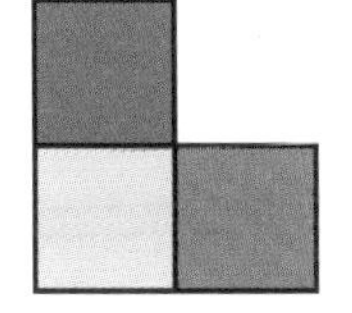 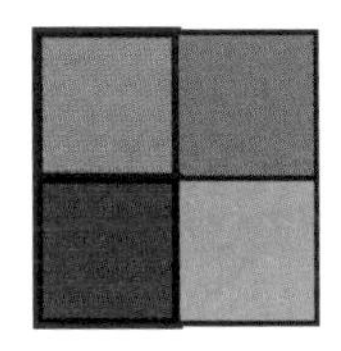 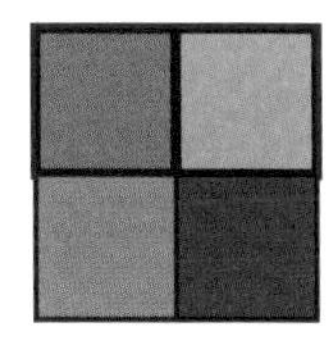 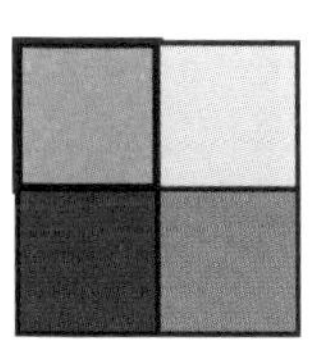 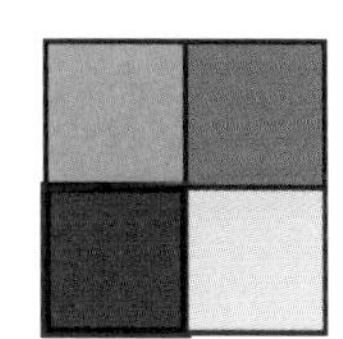 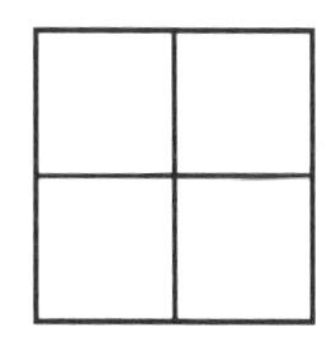

SOLUTION page 127

HOW MANY

How many discrete triangles can be seen in this diagram?

SOLUTION page 132

There are 64 individual squares on a chess board. Excluding these, how many other (combined) squares feature on a chess board? How can you work this out, rather than counting?

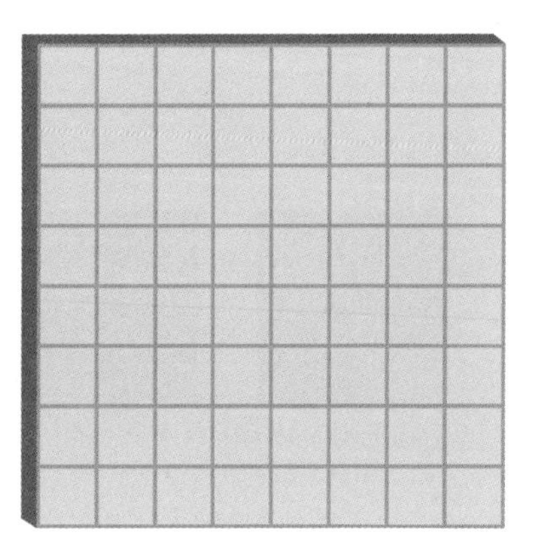

SOLUTION page 127

CUBE NET

How many, and which, of the five cubes shown can be formed from the cube net below?

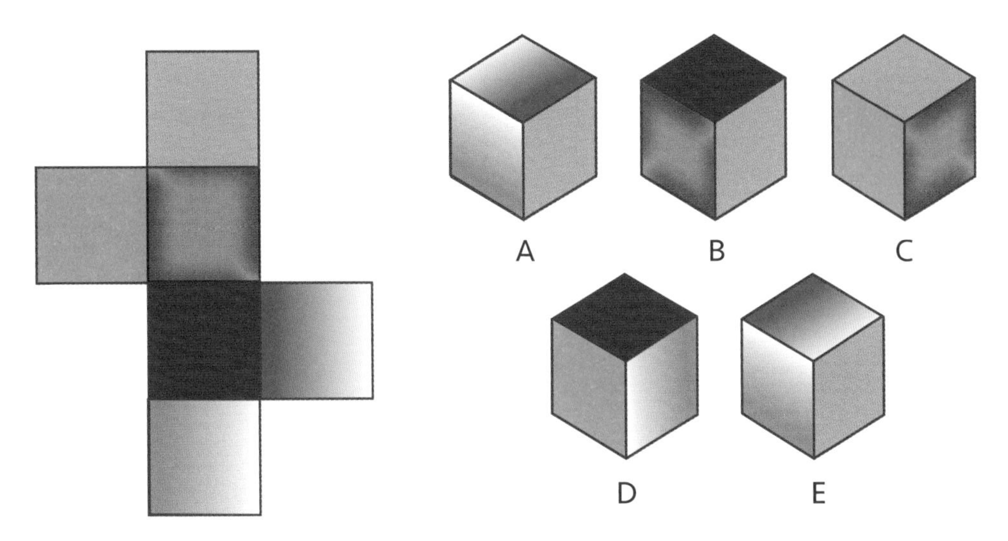

SOLUTION page 127

How many, and which, of the five cubes shown can be formed from the cube net below?

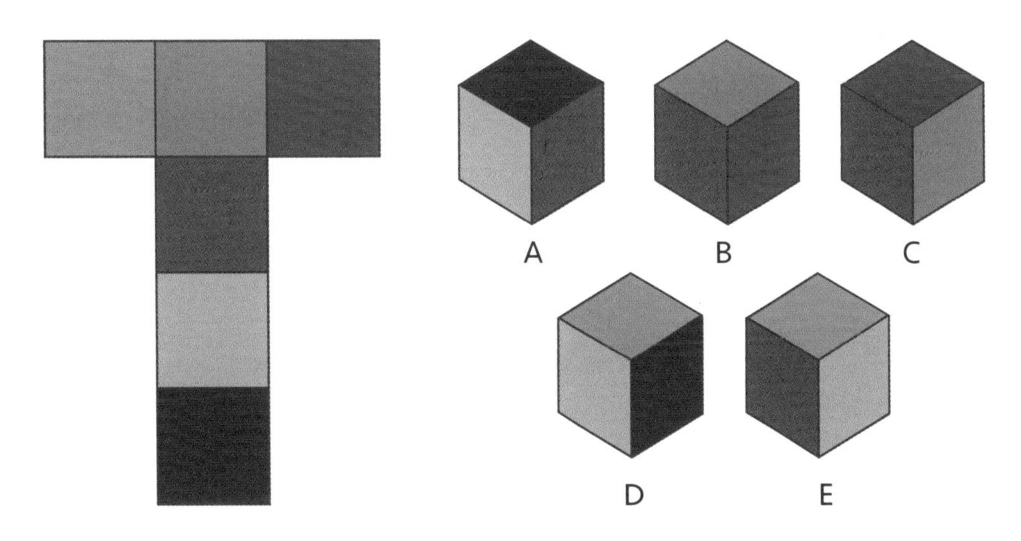

SOLUTION page 140

PERIMETER

Using the information provided, with every internal angle a right angle, what is the total perimeter measurement of this diagram?

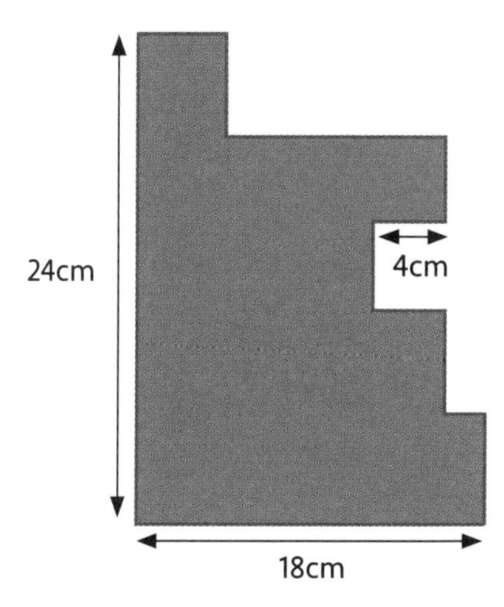

SOLUTION page 132

The letter E is made of five identical blocks. What is its perimeter?

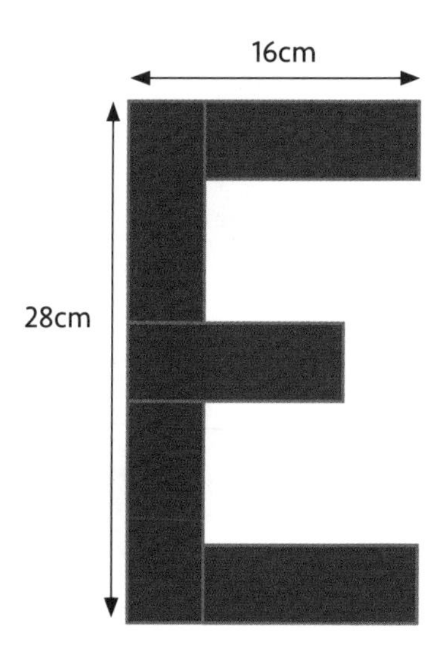

SOLUTION page 126

SHAPE UP

103 If a square sheet of paper is folded along the dotted lines, and cut along the solid lines, which pattern is produced? Cutting is done before folding.

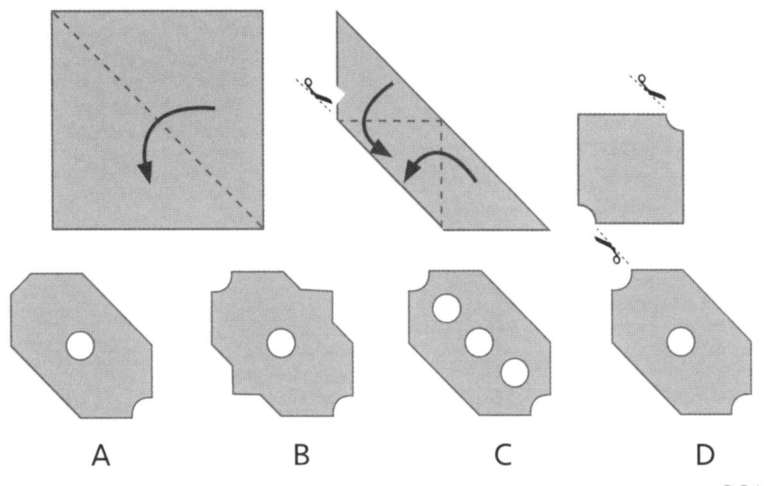

A B C D

SOLUTION page 131

104 If a square sheet of paper is folded along the dotted lines, and cut along the solid lines, which pattern is produced? Cutting is done before folding.

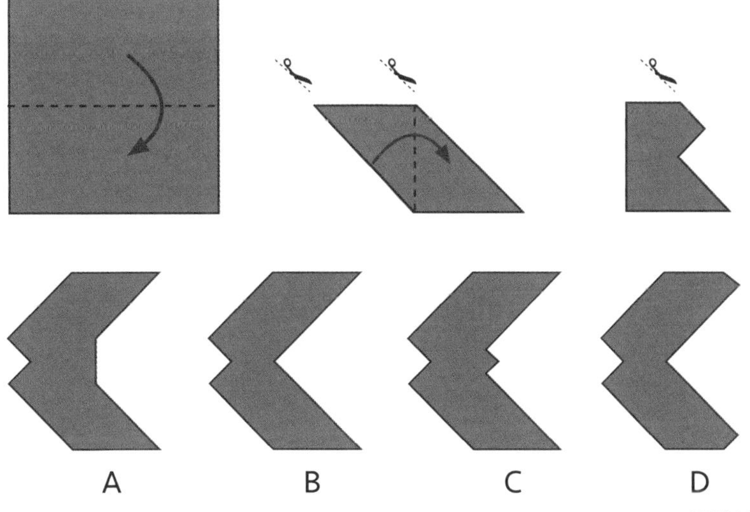

A B C D

SOLUTION page 126

ANGLE GRINDER

105 From the information provided, work out the value of the angle indicated by the question mark.

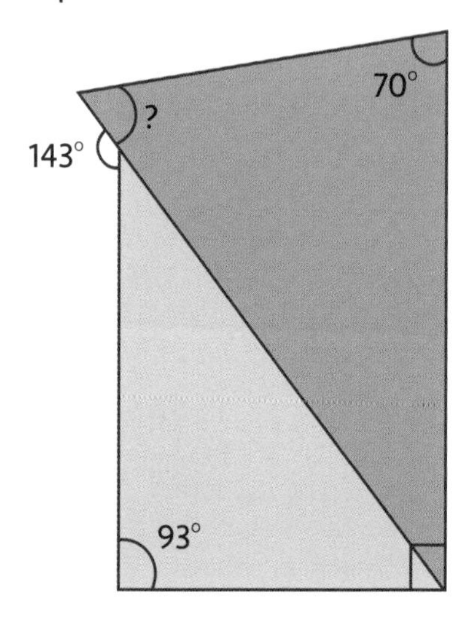

SOLUTION page 127

106 The larger outer square comprises four isosceles triangles, four equilateral triangles and one smaller square set at 45°. What is the value of the smaller angles of the isosceles triangles?

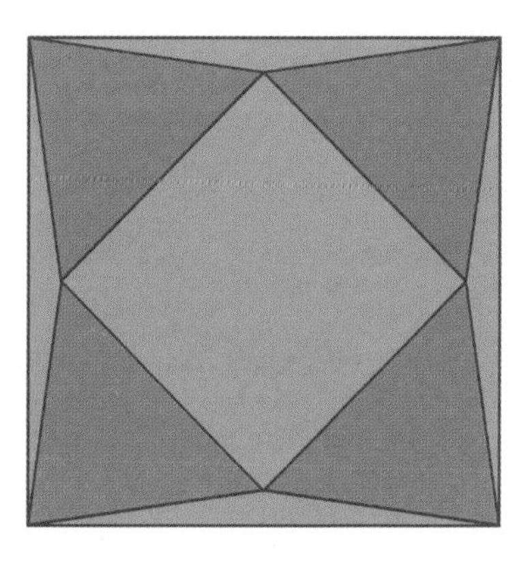

SOLUTION page 128

AREA MAZE (3D)

107 What area is represented by the question mark?

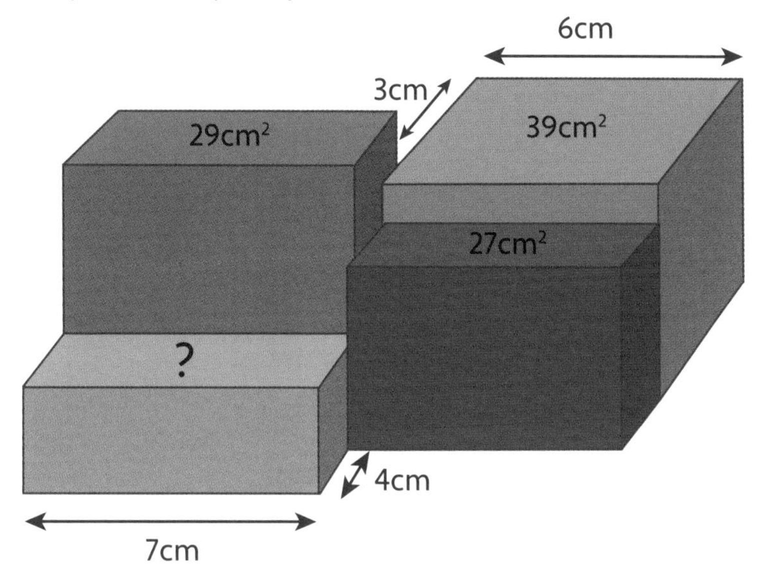

SOLUTION page 128

108 What area is represented by the question mark?

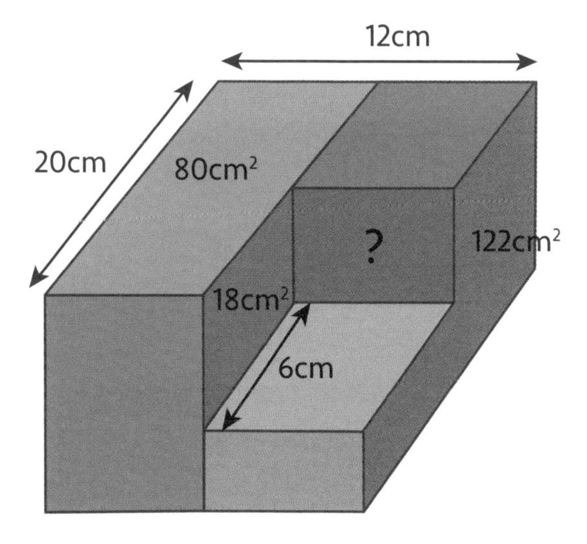

SOLUTION page 128

PENTOMINOES

109 Arrange nine of the twelve pentominoes pictured (as they are and/or rotated and/or reflected) to create the pictured number one.

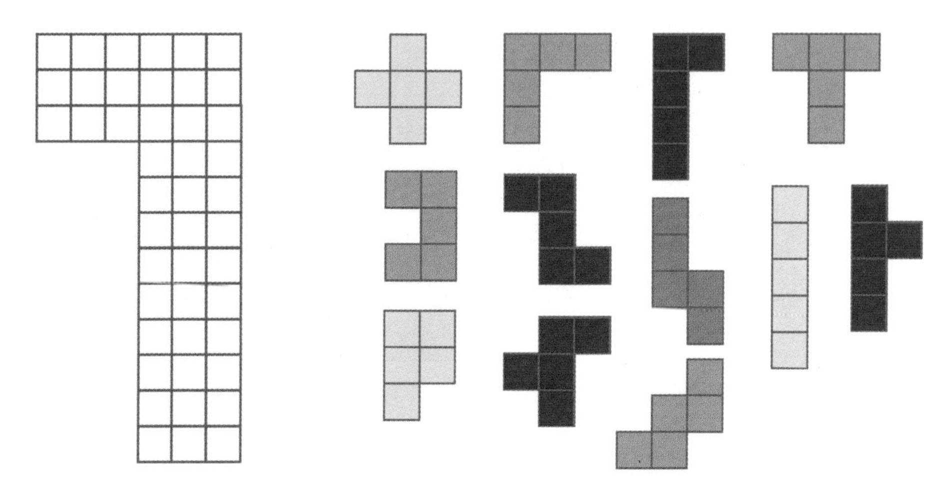

SOLUTION page 128

110 Arrange nine of the twelve pentominoes pictured (as they are and/or rotated and/or reflected) to create the pictured bell.

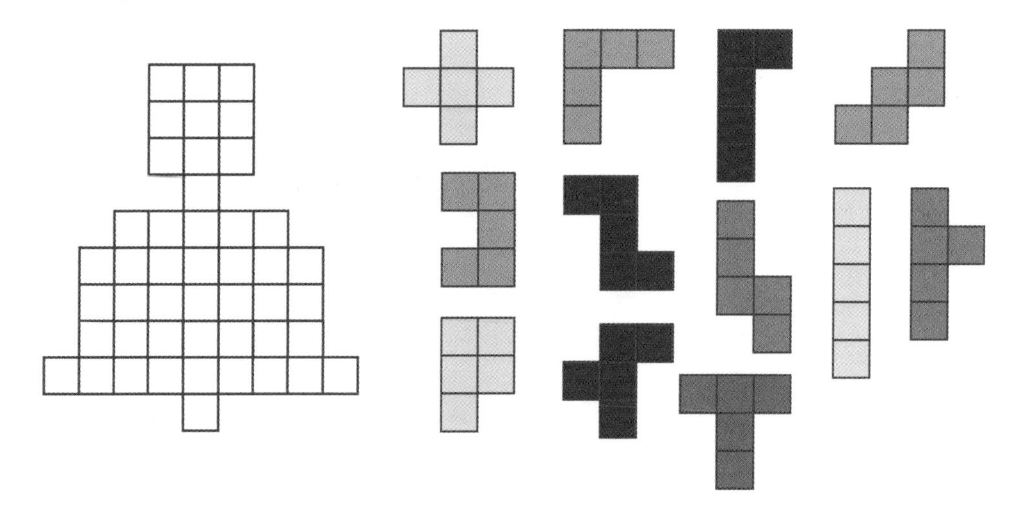

SOLUTION page 128

Card Sharp

Captain Croupier posed a challenge to the semi-finalists, including you, in the casino poker contest, with the first semi-finalist to answer correctly getting a free pass to the final. Arrange the thirteen cards in a suit in twelve rows of three cards each.

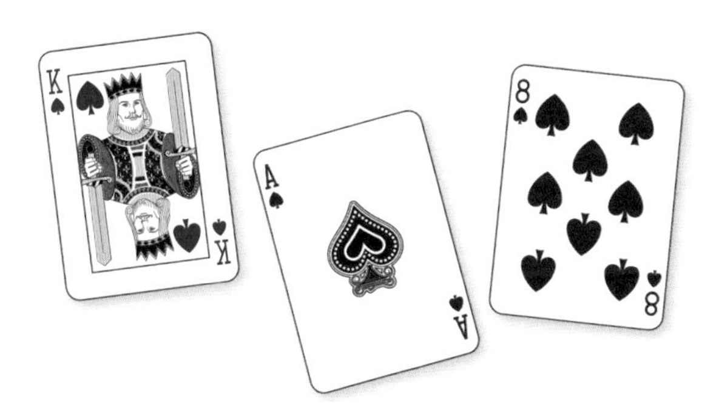

SOLUTION page 129

Rabbiting On

Each year the local primary school has its own rabbits. When they are all out on the field, they are kept apart – for obvious reasons – with six equal pens being arranged using 24 fence panels. However, twelve fence panels were damaged in a storm, and now there are only 12 fence panels available. How can all six rabbits be placed on the field at the same time in equal-sized pens?

SOLUTION page 129

LINE DRAWING

113 Use six lines, running from one side to another, to divide the grid into sixteen areas, with each area containing one dot.

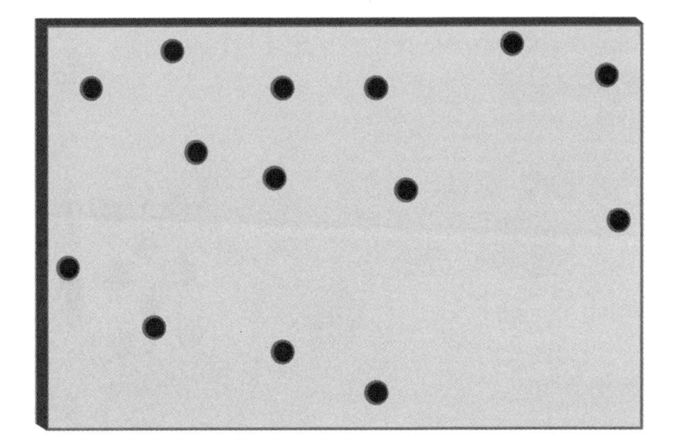

SOLUTION page 129

114 Use three lines, running from one side to another, to divide the grid into five areas. Each area must contain one square and two circles.

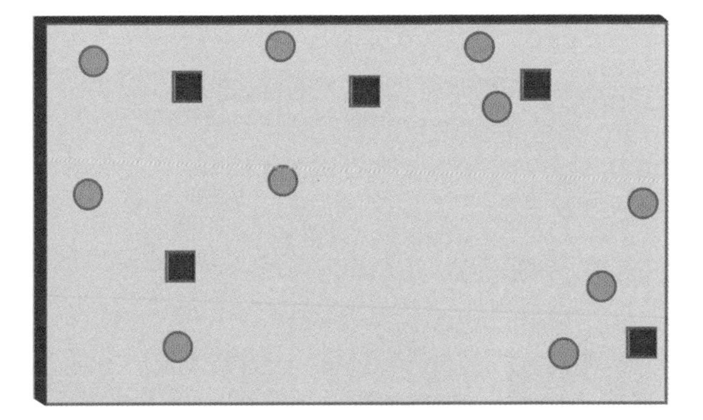

SOLUTION page 125

AREA MAZE

What is the area represented by the question mark?

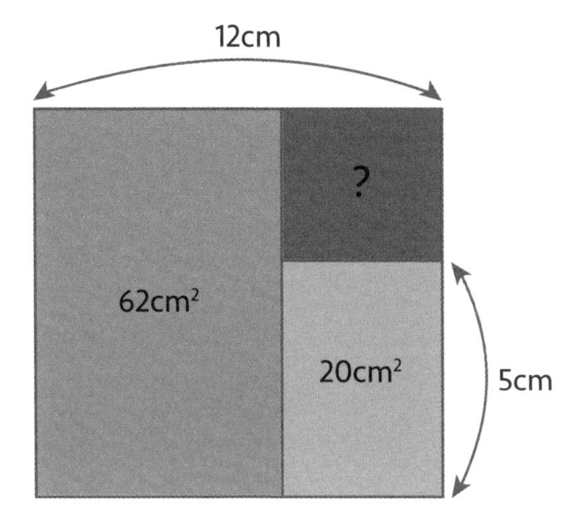

SOLUTION page 135

What is the length represented by the question mark?

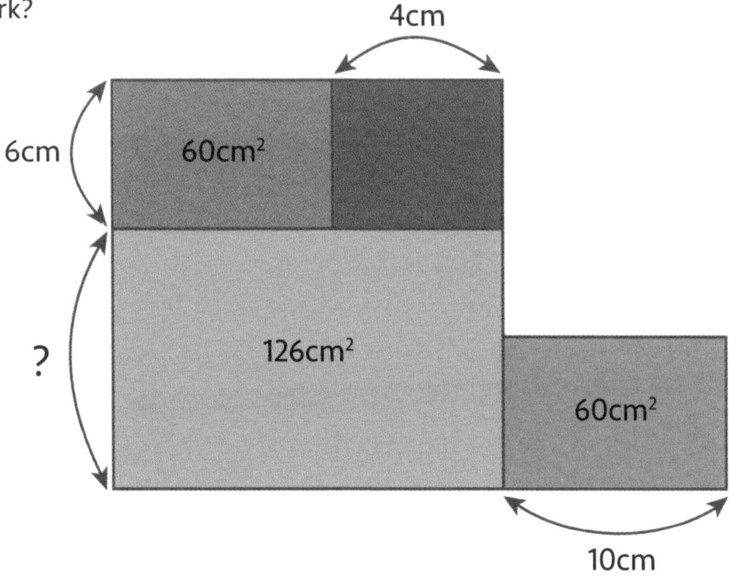

SOLUTION page 129

AREA MAZE (PERCENTAGES)

117

To the nearest percent, the yellow area is how much of the whole shape?

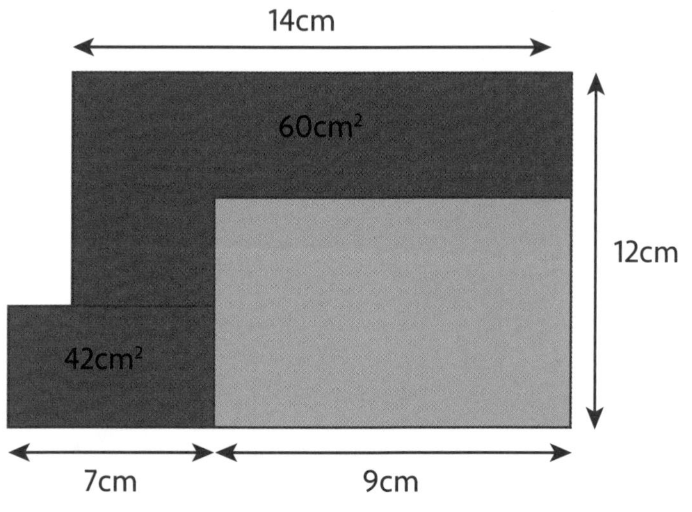

SOLUTION page 123

118

To the nearest percent, the yellow area is how much of the whole shape?

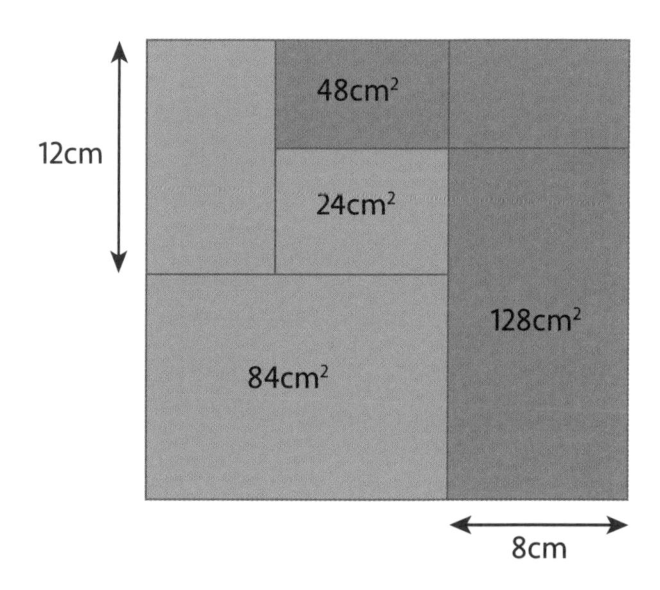

SOLUTION page 130

SPLIT DECISION

Split the grid into three identical shapes, following the marked lines. Shapes may be rotated, but not flipped.

SOLUTION page 118

Split the grid into three identical shapes, following the marked lines. Shapes may be rotated, but not flipped.

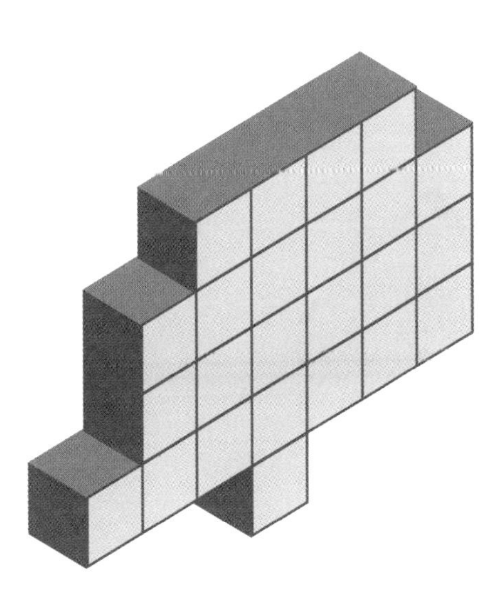

SOLUTION page 130

TANGRAM

121 The seven different shapes provided can be combined to form the outlined Christmas tree. How?

SOLUTION page 130

122 The seven different shapes provided can be combined to form the outlined factory with smokestack. How?

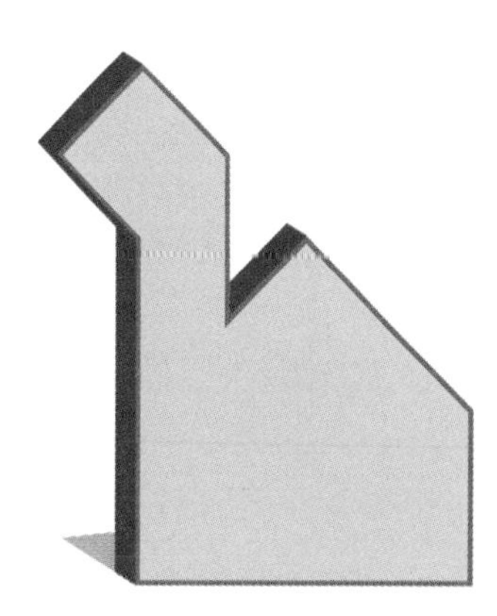

SOLUTION page 124

SQUARE DEAL

Divide this shape into two equal halves, marking the point of rotation.

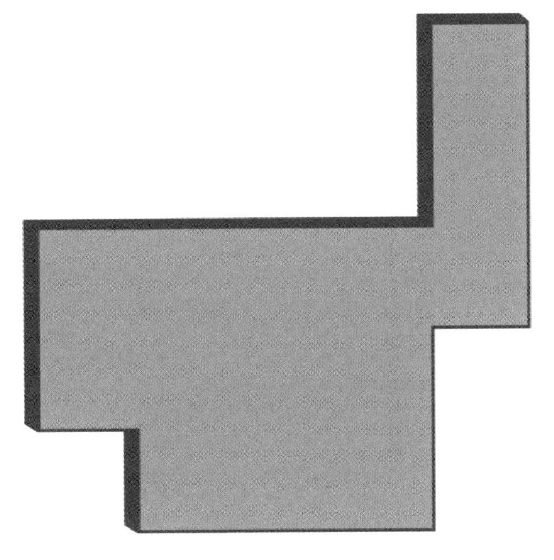

SOLUTION page 127

Split the shape into two pieces that can be arranged to form an 8x8 square.

SOLUTION page 131

MATCHSTICKS

Add four matches to divide the square into two equal parts.

SOLUTION page 125

A) Remove three matches to leave three squares.
B) Remove three matches to leave five squares.
C) Remove five matches to leave two squares.

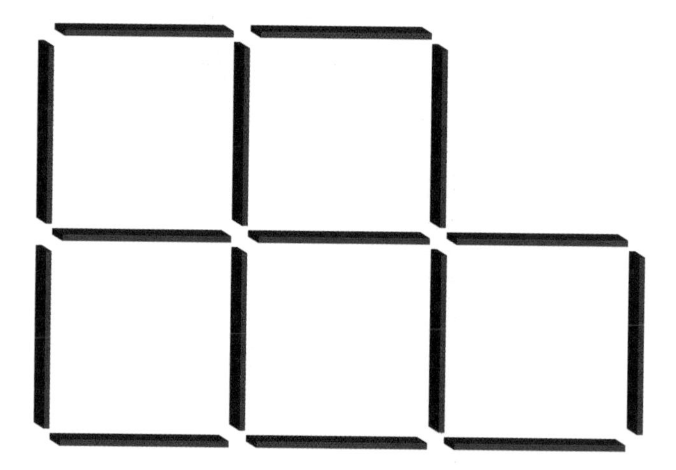

SOLUTION page 131

ADDING LINES

127 Add two squares to the design to add ten more squares to the symmetrical pattern.

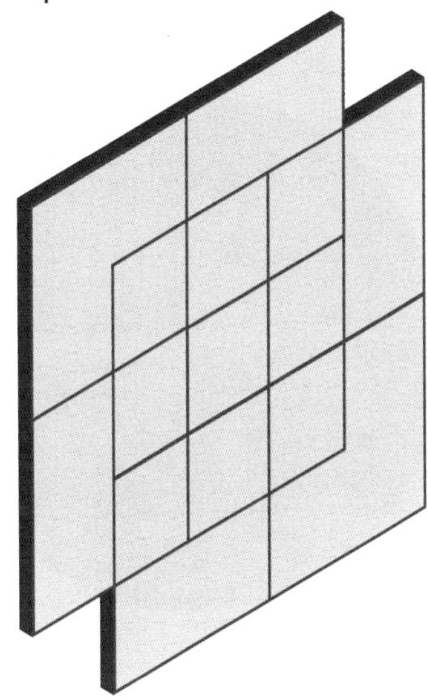

SOLUTION page 131

128 Add just one rectangle to the design to add 17 more rectangles, including one square, to the pattern.

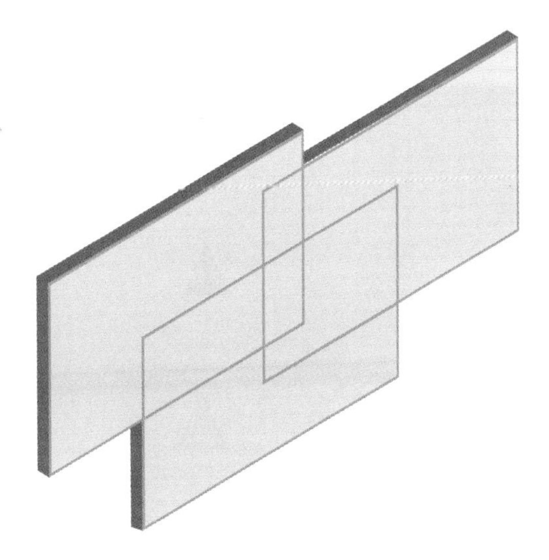

SOLUTION page 138

LINE DRAWING

129 Use four lines, running from one side to another, to divide the circle into eleven areas.

SOLUTION page 131

130 Use three lines, running from one side to another, to divide the rectangle into seven areas. Each area must contain one triangle.

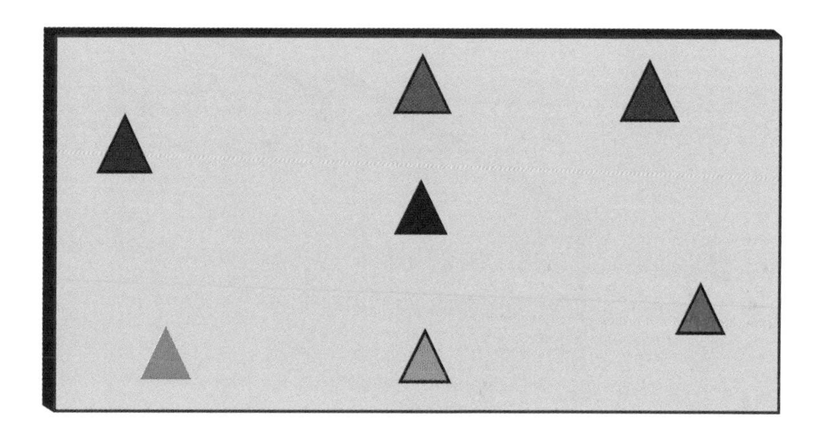

SOLUTION page 141

CONGRUENT SHAPES

131 Split the diagram into two congruent shapes, identical when rotated and/or flipped. There are no grid lines to follow, but there is enough information in the diagram.

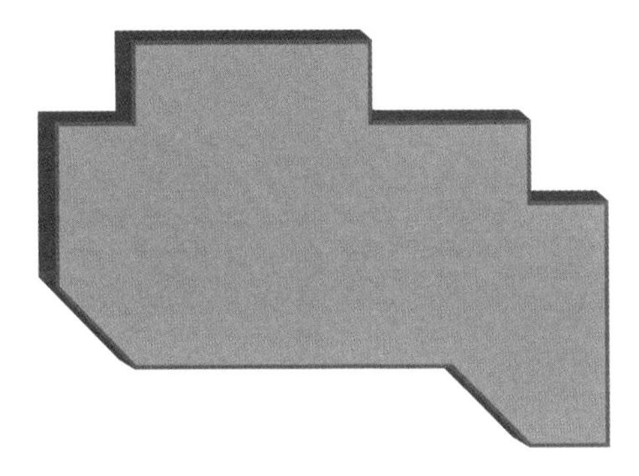

SOLUTION page 128

132 Split the grid into three congruent shapes, identical when rotated and/or flipped.

SOLUTION page 132

XOXO

133 Divide the grid into four congruent shapes, with the point of rotation at the centre of the grid. We have included blue, red and yellow squares as clues – no shape contains more than one of each.

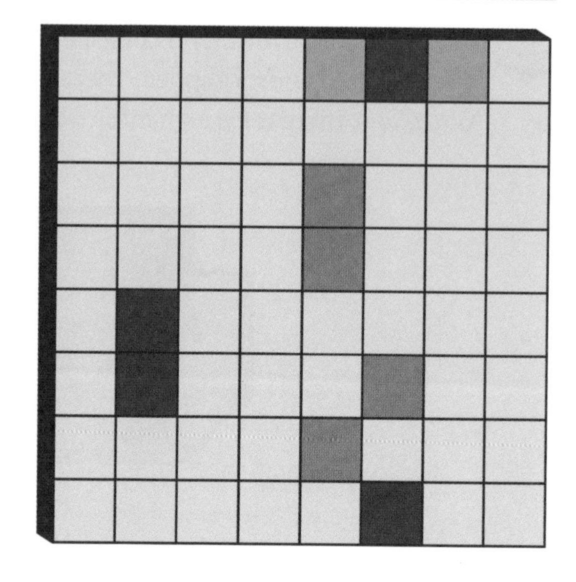

SOLUTION page 126

134 Divide the grid into four congruent shapes, with the point of rotation at the centre of the grid. We have included four blue squares and four red squares as clues – no shape contains more than one blue square and one red square.

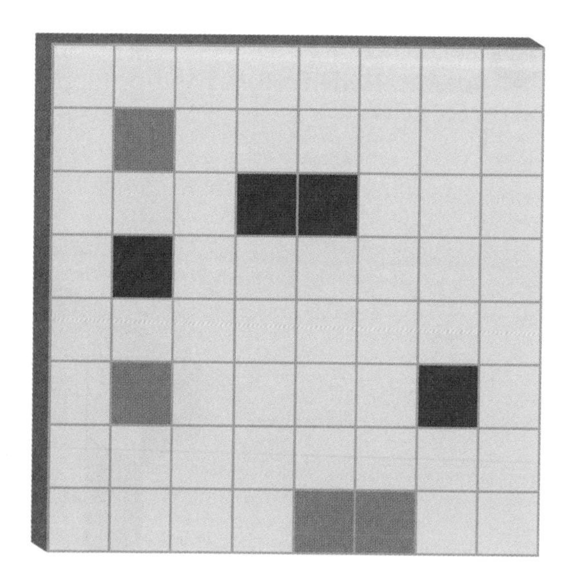

SOLUTION page 132

OVERLAID

What is the smallest number of overlaid paper squares that can be used to create this pattern? In what order were they laid?

SOLUTION page 139

136 What is the smallest number of overlaid paper squares that can be used to create this pattern? In what order were they laid?

SOLUTION page 132

OVERVIEW

137 A 2x2x2 cube is made of eight differently coloured cubes of the same size. Two of the cubes have been removed. Five of the six faces are shown, as seen from above. A thicker border indicates that this cube is on the bottom row. When looking at the sixth and final face, as seen from above, what do you see?

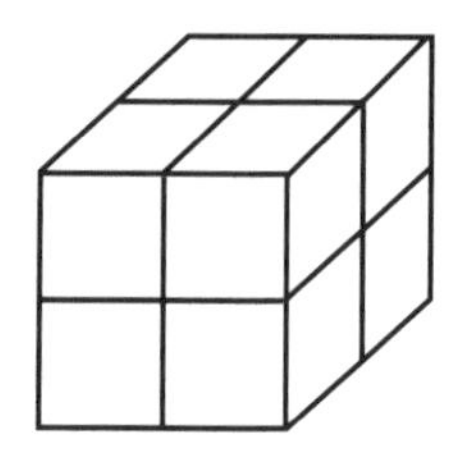

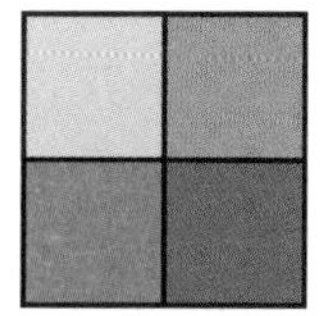 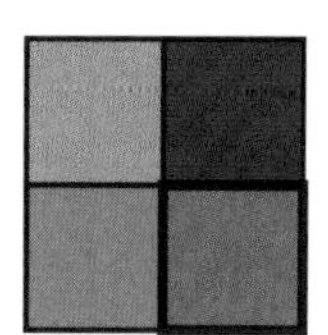 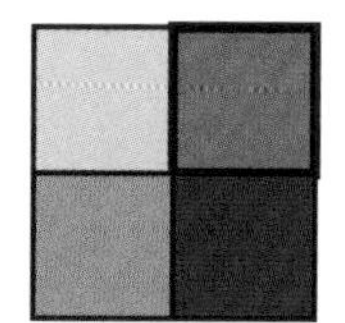 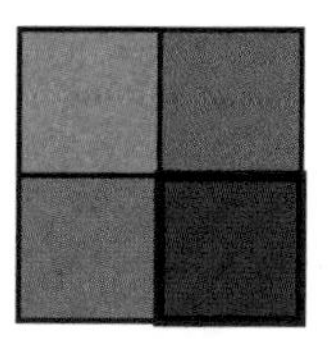 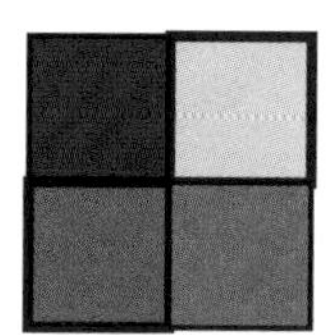 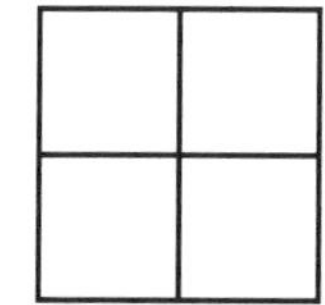

SOLUTION page 139

138 A 2x2x2 cube is made of eight differently coloured cubes of the same size. Four of the cubes have been removed. Five of the six faces are shown, as seen from above. A thicker border indicates that this cube is on the bottom row. When looking at the sixth and final face, as seen from above, what do you see?

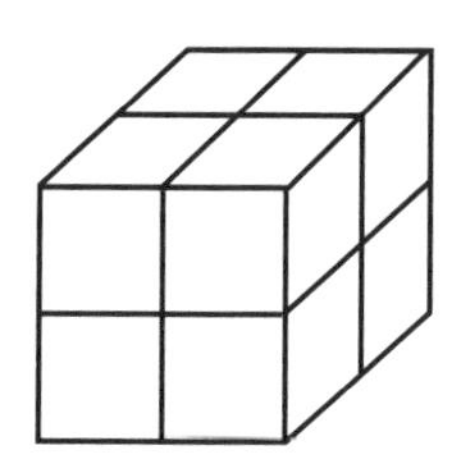

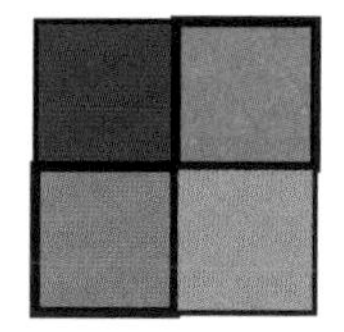 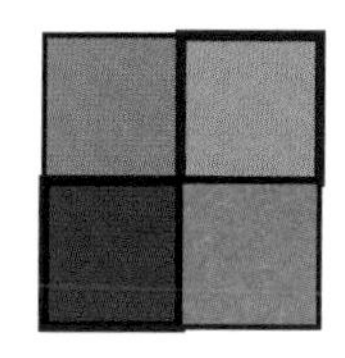 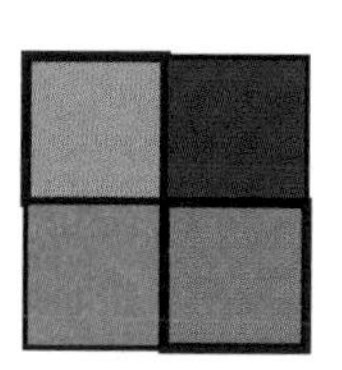 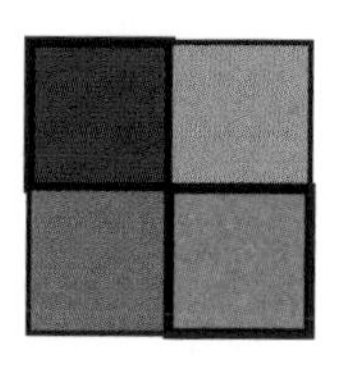 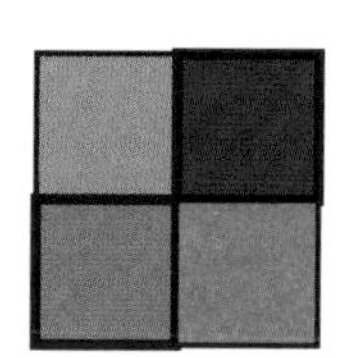 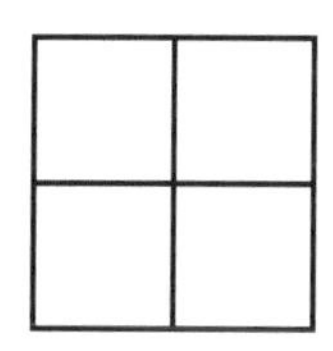

SOLUTION page 132

HOW MANY

139 How many discrete triangles can be seen in this diagram?

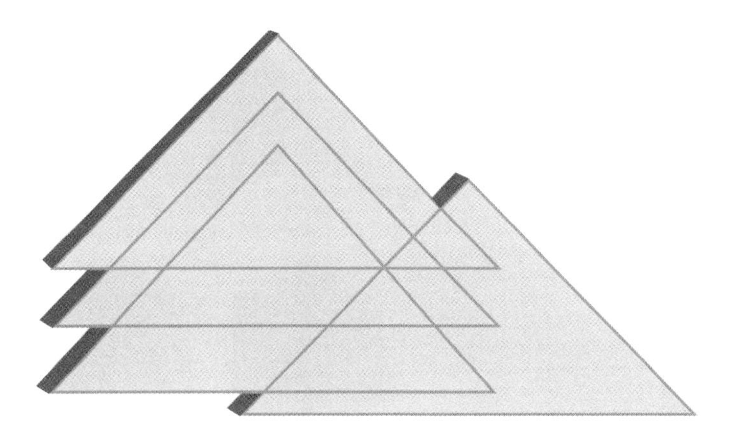

SOLUTION page 136

140 How many discrete squares can be seen in this diagram?

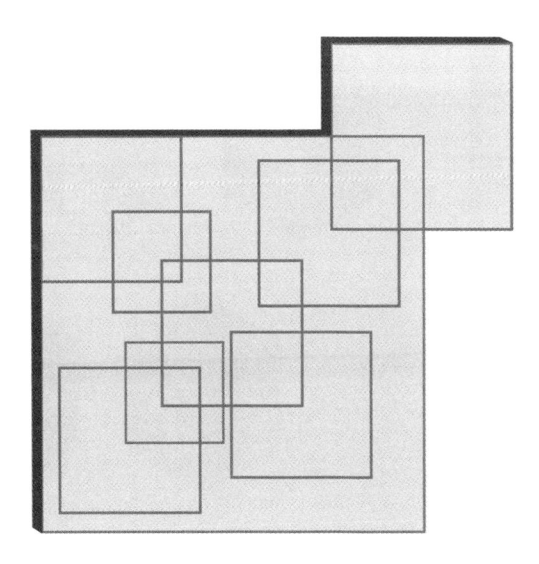

SOLUTION page 133

CUBE NET

141 How many, and which, of the five cubes shown can be formed from the cube net below?

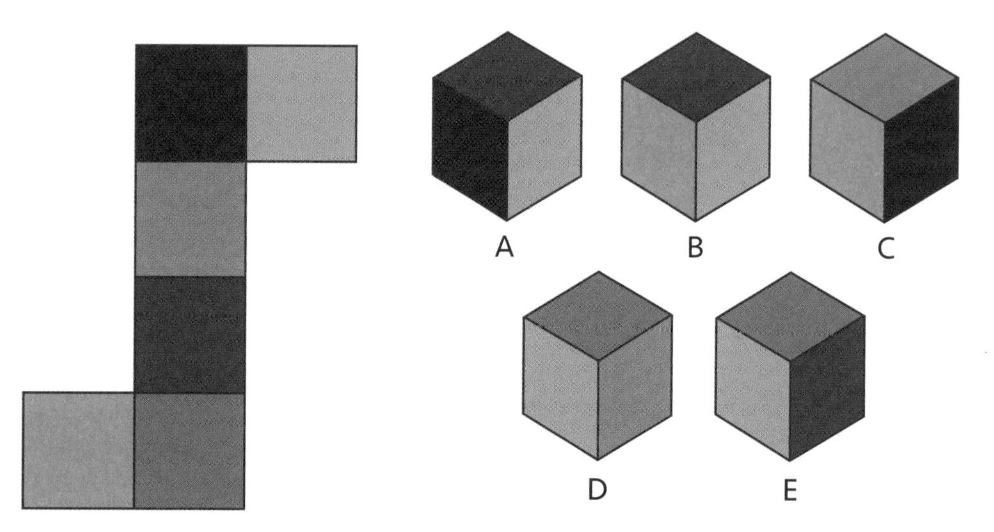

SOLUTION page 127

142 How many, and which, of the five cubes shown can be formed from the cube net below?

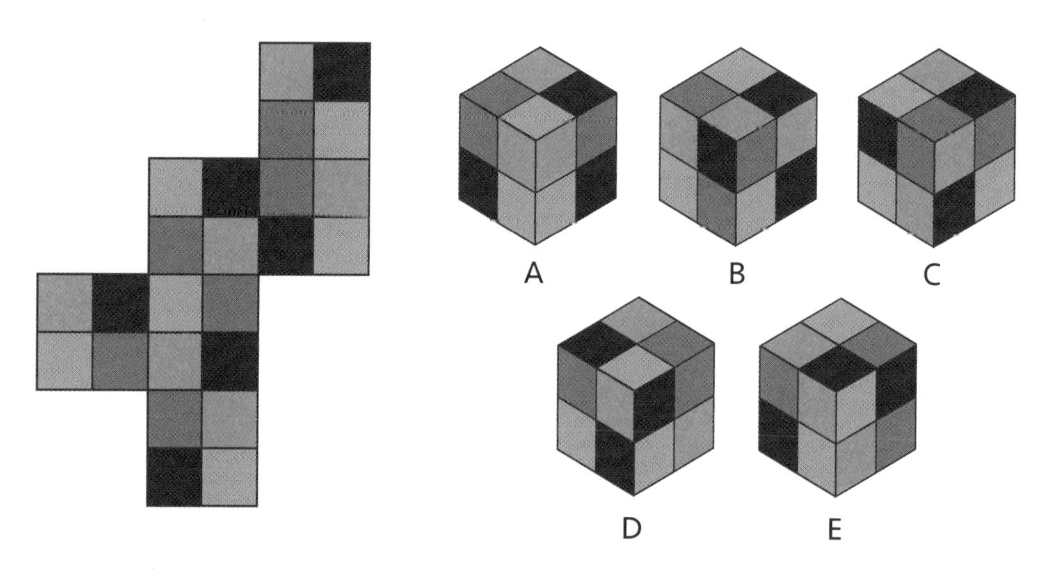

SOLUTION page 133

PERIMETER

143 Each block in the parquet flooring is twice as long as wide. If the overall perimeter of this area is 486cm, how wide is each block?

SOLUTION page 135

144 Using the information provided, what is the total perimeter measurement of this stylised number 7?

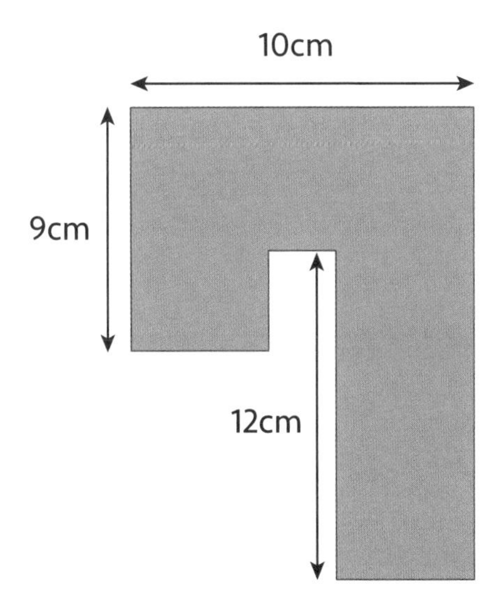

SOLUTION page 141

SHAPE UP

145 If a square sheet of paper is folded along the dotted lines, and cut along the solid lines, which pattern is produced? Cutting is done before folding.

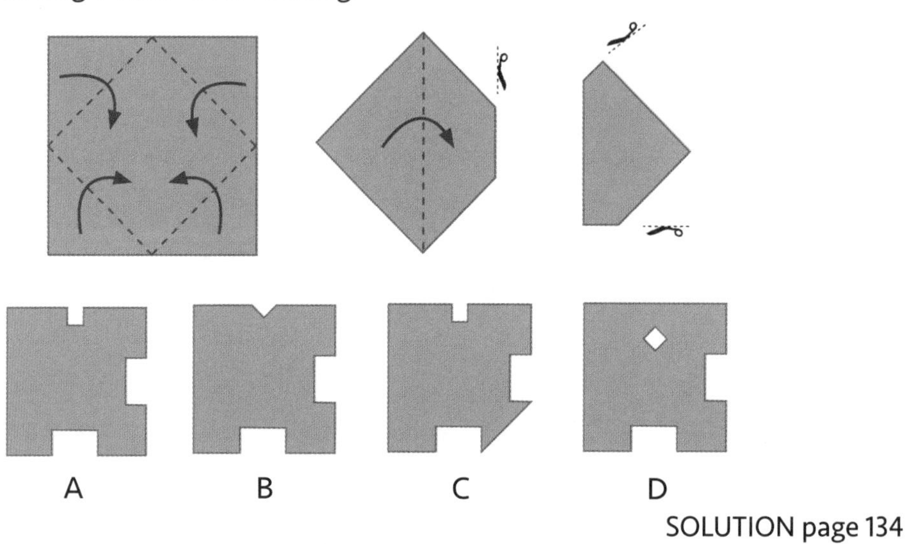

SOLUTION page 134

146 If a triangular sheet of paper is folded along the dotted lines, and cut along the solid lines, which pattern is produced?

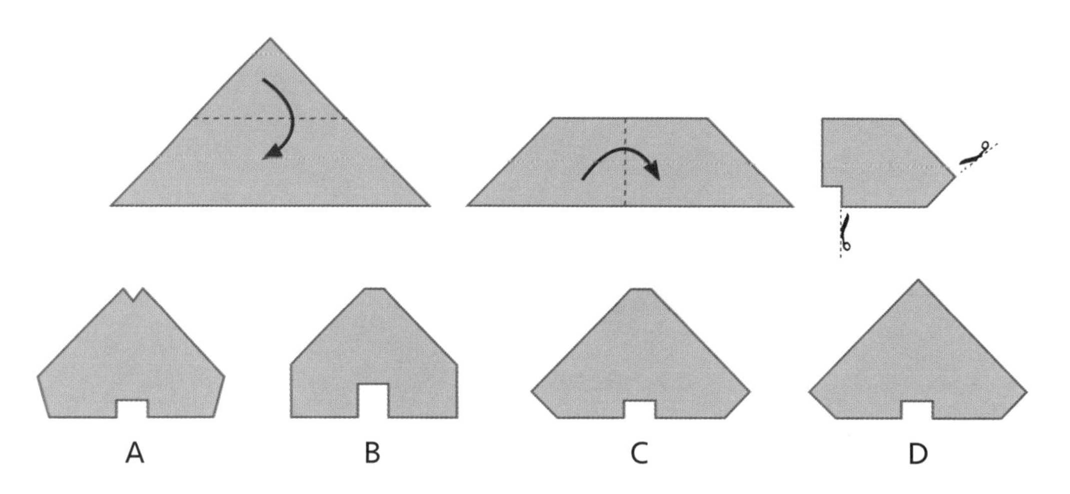

SOLUTION page 140

ANGLE GRINDER

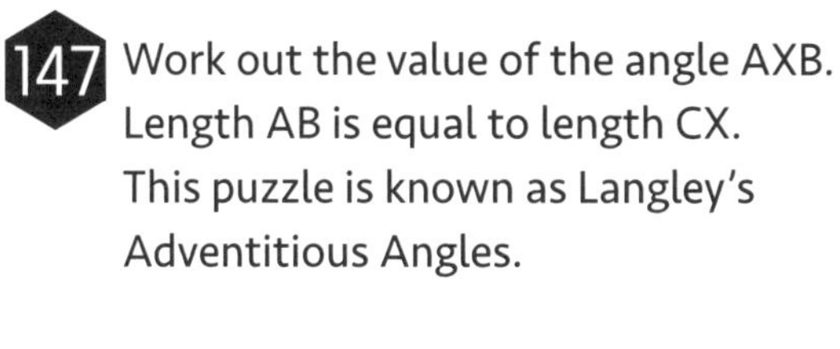

147 Work out the value of the angle AXB. Length AB is equal to length CX. This puzzle is known as Langley's Adventitious Angles.

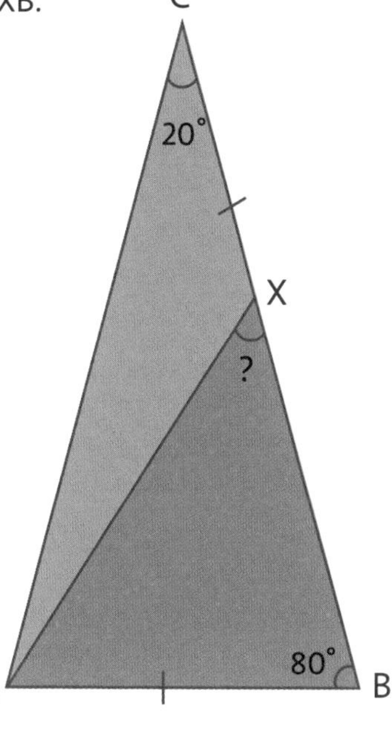

SOLUTION page 134

148 What is the value of angle GHJ?

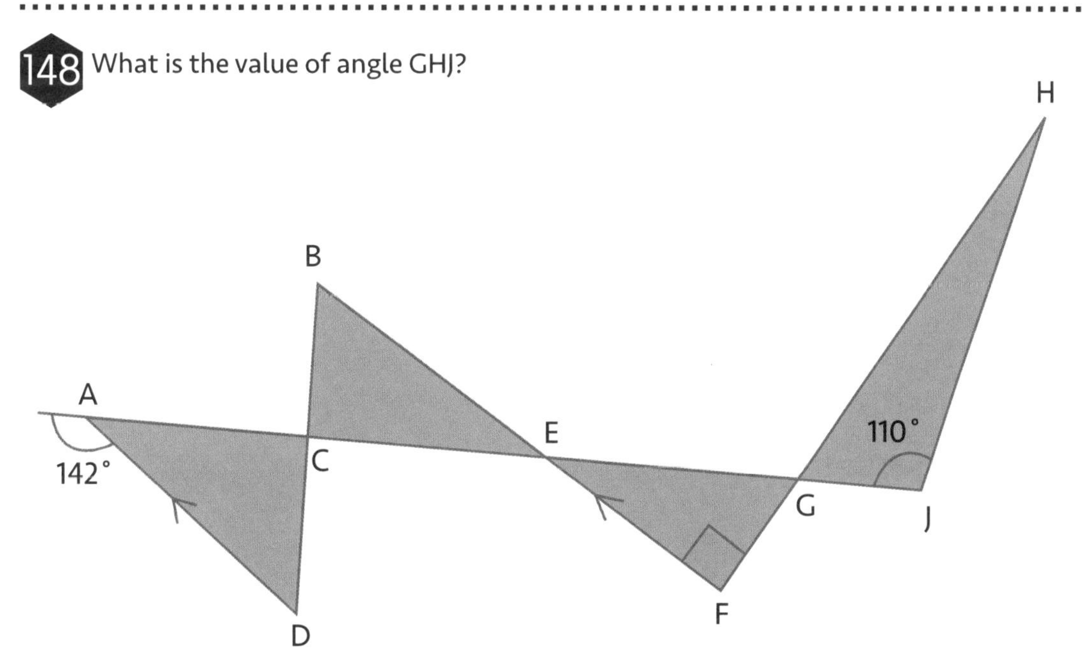

SOLUTION page 133

AREA MAZE (3D)

149 What area is represented by the question mark?

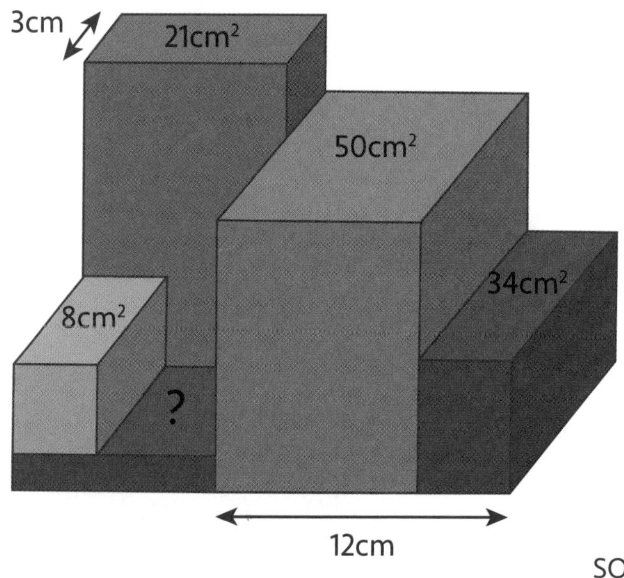

SOLUTION page 134

150 What length is represented by the question mark?

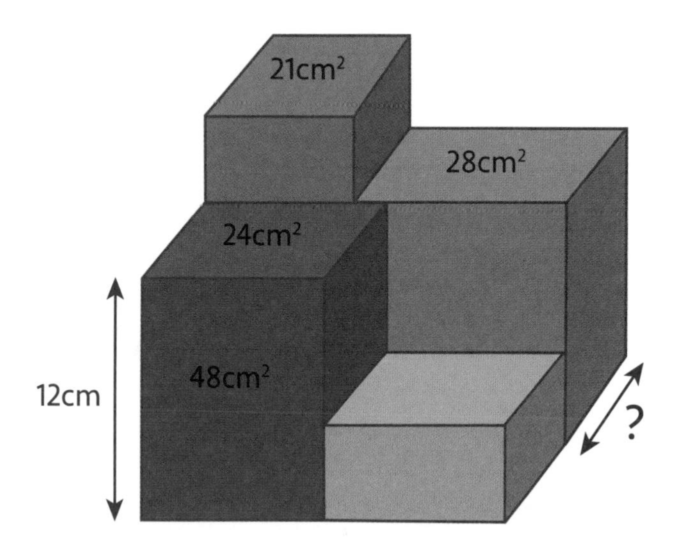

SOLUTION page 141

PENTOMINOES

151 Arrange nine of the twelve pentominoes pictured (as they are and/or rotated and/or reflected) to create the pictured bug face.

SOLUTION page 138

152 Arrange all twelve pentominoes pictured (as they are and/or rotated and/or reflected) to create the pictured cat.

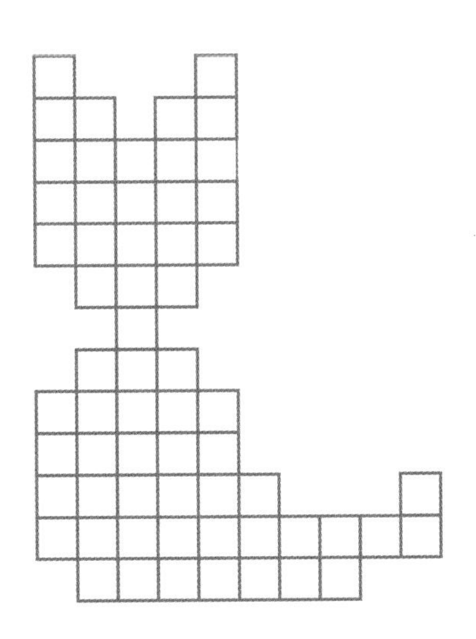

SOLUTION page 135

ARRANGEMENTS

153 Coining it in

With ten discs arranged to form a five-disc row and a six-disc row, move one disc to form two six-disc rows.

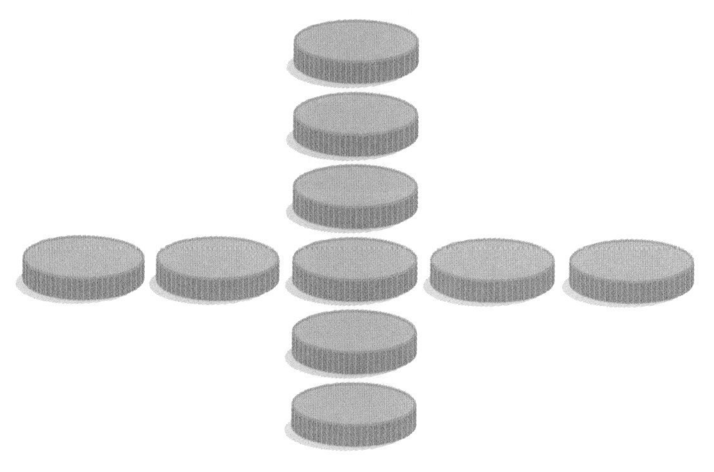

SOLUTION page 142

154 Coining it in

Ten discs are arranged to form an equilateral triangle with the vertex at the bottom. Move just three discs to form an equilateral triangle with the vertex at the top.

SOLUTION page 133

AREA MAZES

155 What is the length represented by the question mark?

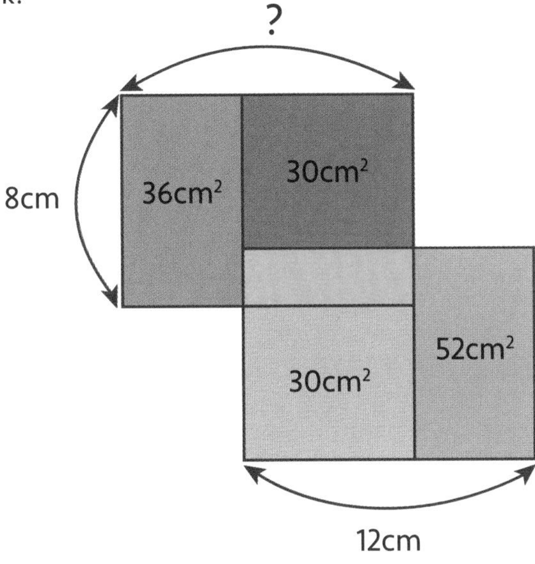

SOLUTION page 142

156 What is the area represented by the question mark?

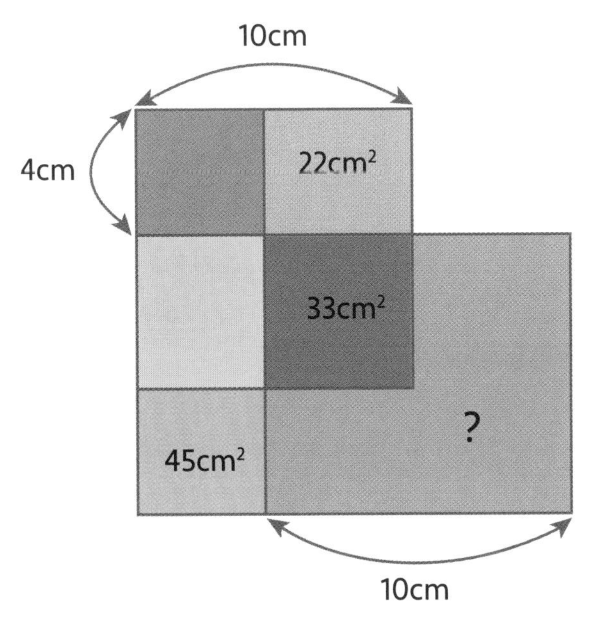

SOLUTION page 135

AREA MAZE (PERCENTAGES)

157 As a fraction, what proportion of the whole shape is the blue area?

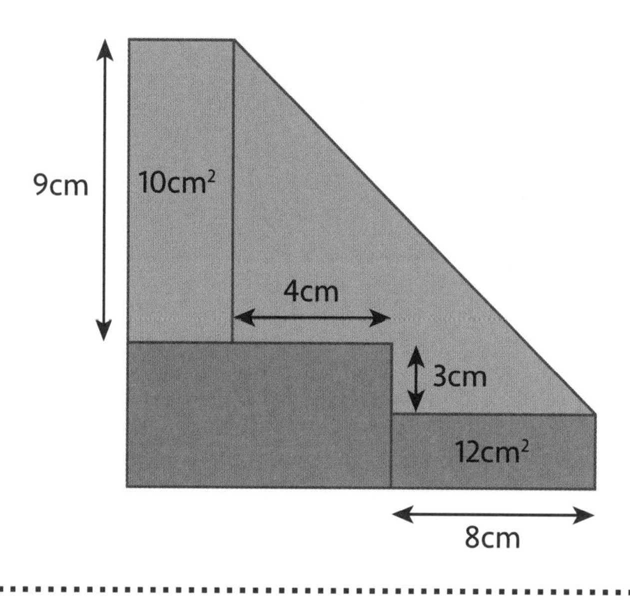

SOLUTION page 136

158 To the nearest percent, the yellow area is how much of the whole shape?

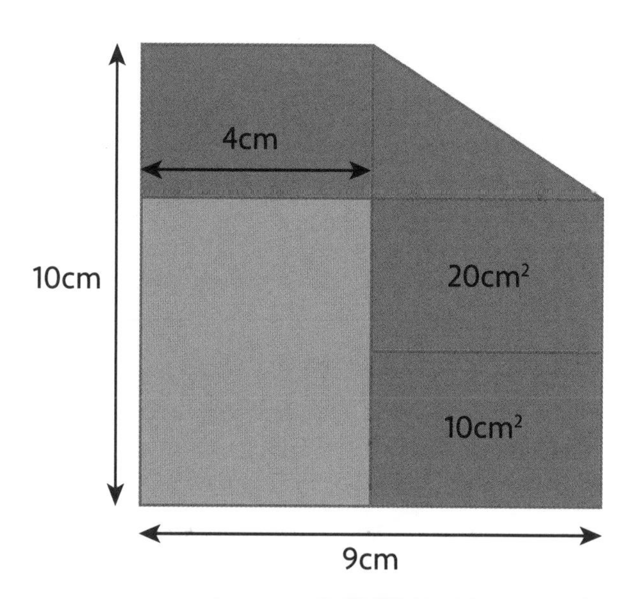

SOLUTION page 143

SPLIT DECISION

Split the grid into five identical shapes, following the marked lines. Shapes may be rotated, but not flipped.

SOLUTION page 136

Split the grid into three identical shapes, following the marked lines. Shapes may be rotated, but not flipped.

SOLUTION page 136

TANGRAM

161 The seven different shapes provided can be combined to form the outlined letter C. How?

SOLUTION page 136

162 The seven different shapes provided can be combined to form the outlined arrow. How?

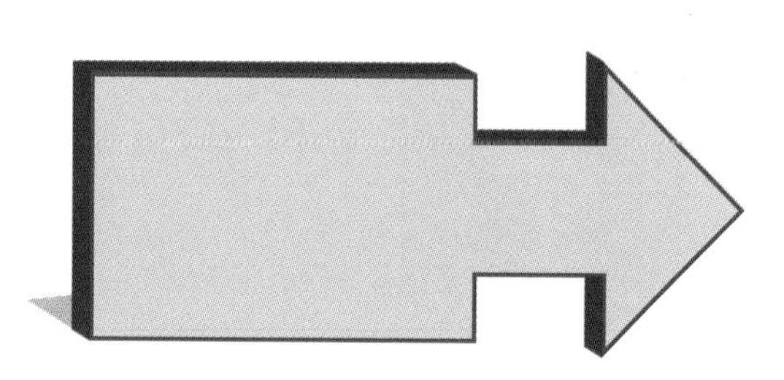

SOLUTION page 133

SQUARE DEAL

163 Split the shape into two pieces that can be arranged to form an 8x8 square.

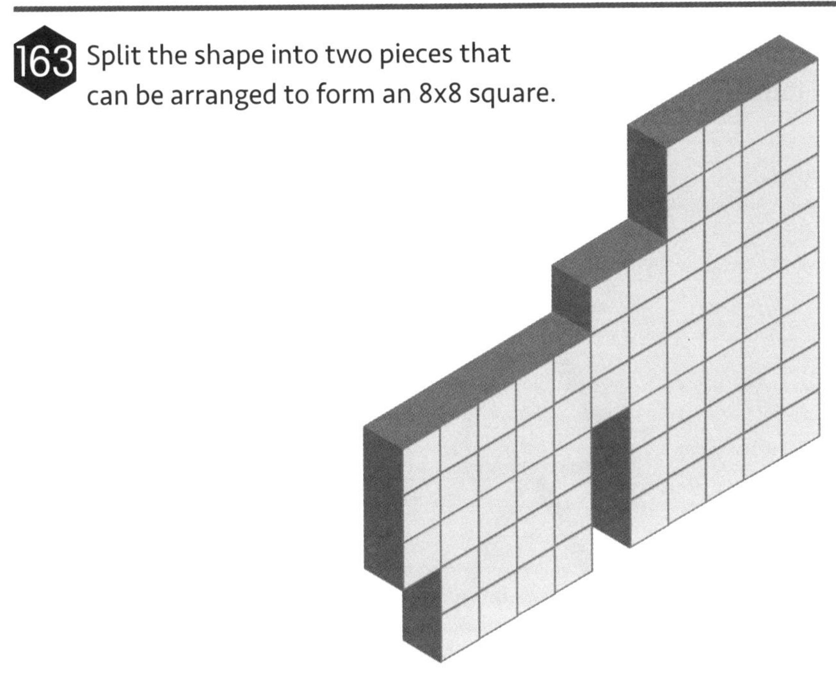

SOLUTION page 137

164 Divide this quadrilateral into two equal halves with just one line.

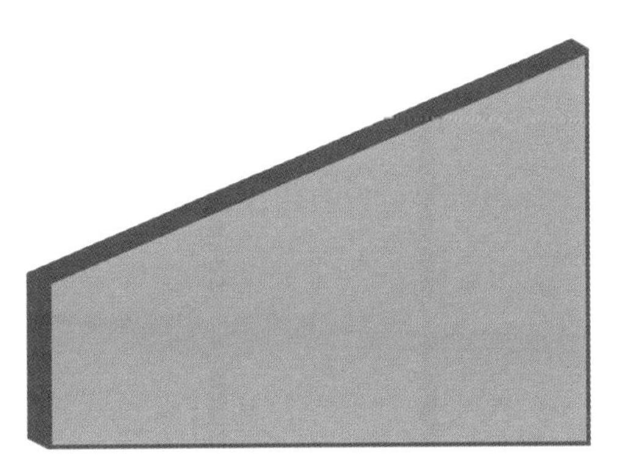

SOLUTION page 135

MATCHSTICKS

Move three matches to make four equilateral triangles with no overlapping.

SOLUTION page 131

A) Move four matches to make three squares. Each match must form part of a square, and no matches can overlap.
B) Move three matches to leave three squares. Each match must form part of a square, and no matches can overlap.

SOLUTION page 137

ADDING LINES

167 Add just one similar triangle to the design to add 21 triangles in the pattern.

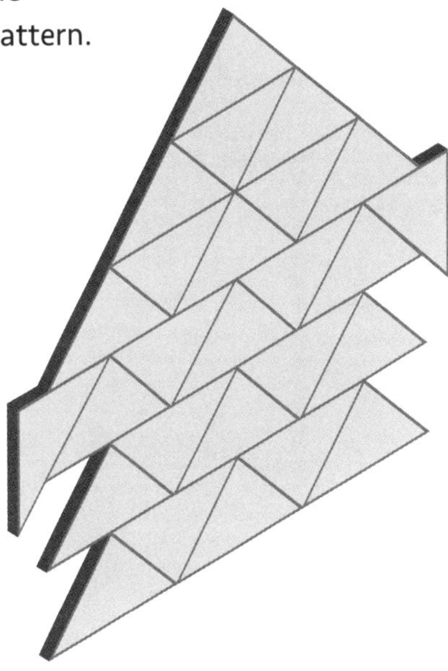

SOLUTION page 137

168 Add just one straight line to the design to create 8 triangles in the pattern.

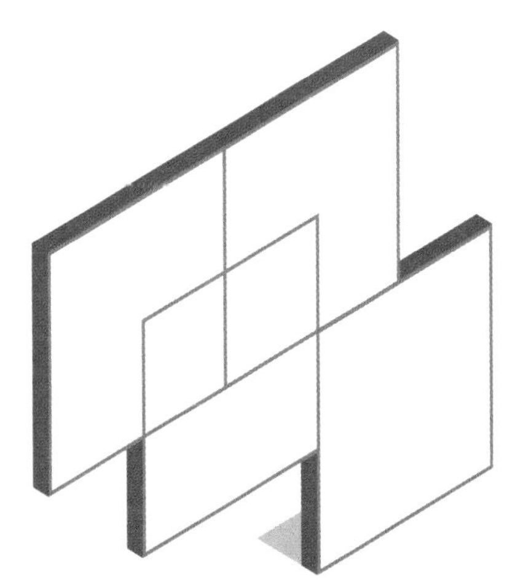

SOLUTION page 131

LINE DRAWING

169 Use five lines, running from one side to another, to divide the square into thirteen areas. Each area must contain one diamond.

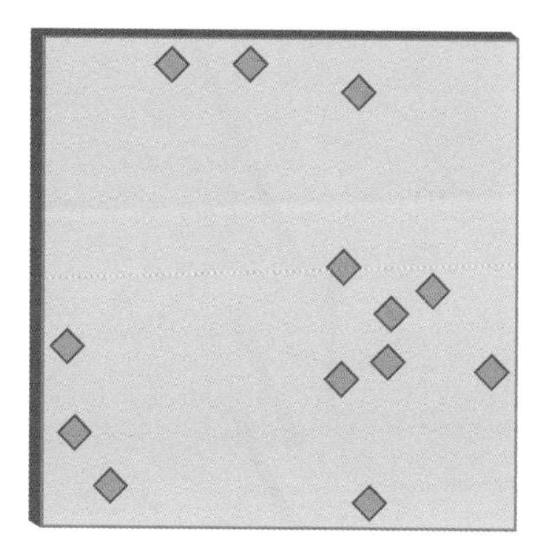

SOLUTION page 138

170 Use four lines, running from one side to another, to divide the rectangle into nine areas. Each area must contain one each of the red, yellow and orange dots.

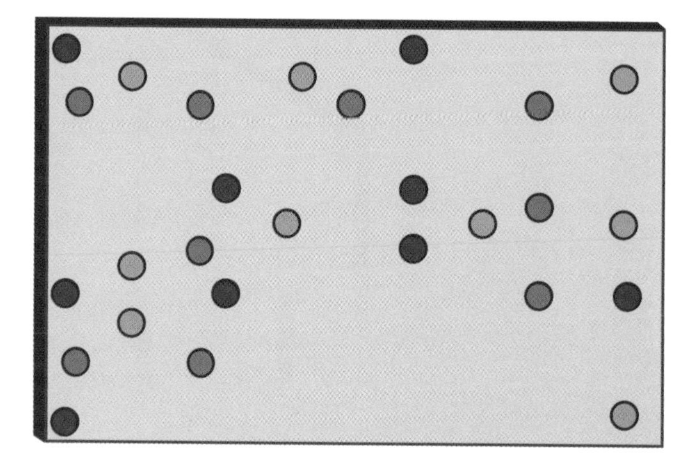

SOLUTION page 131

CONGRUENT SHAPES

171 Split the grid into two congruent shapes, identical when rotated and/or flipped.

SOLUTION page 138

172 Split the grid into two congruent shapes, identical when rotated and/or flipped.

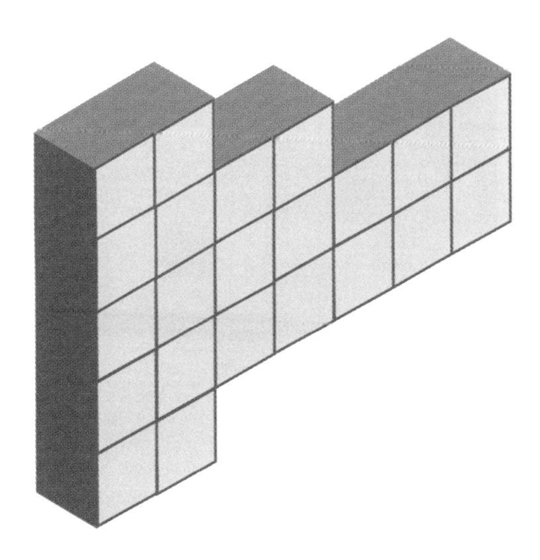

SOLUTION page 138

XOXO

173 Divide the grid, following the lines and with the point of rotation at the centre of the grid, into four matching shapes that each contain one red square. The shapes may be rotated, but not flipped.

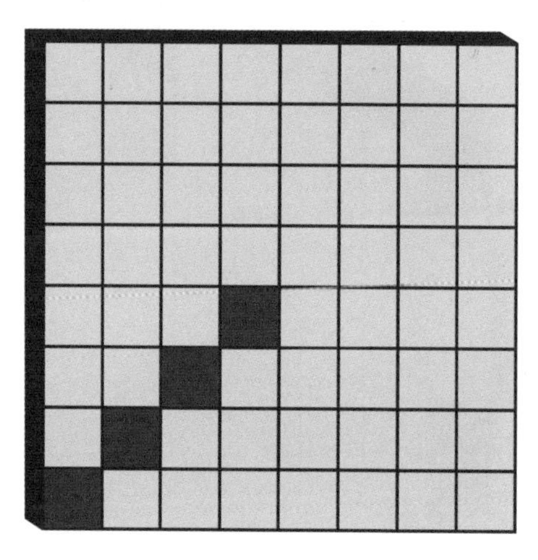

SOLUTION page 138

174 Divide the grid into four congruent shapes, with the point of rotation at the centre of the grid. We have included blue, red and yellow squares as clues – no shape contains more than one of each.

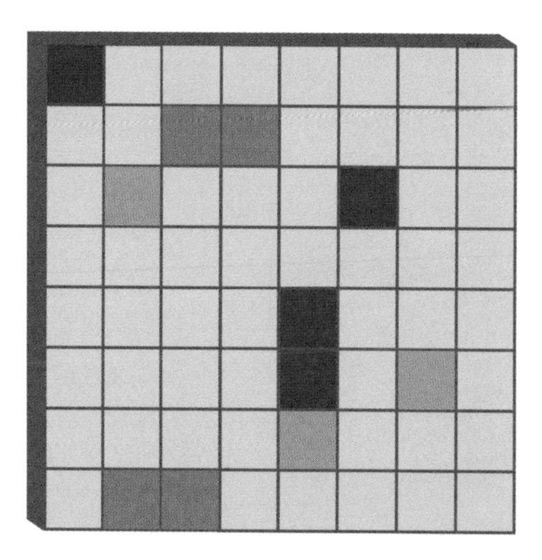

SOLUTION page 139

OVERLAID

175 What is the smallest number of overlaid paper squares that can be used to create this pattern? In what order were they laid?

SOLUTION page 139

176 What is the smallest number of overlaid paper squares that can be used to create this pattern? In what order were they laid?

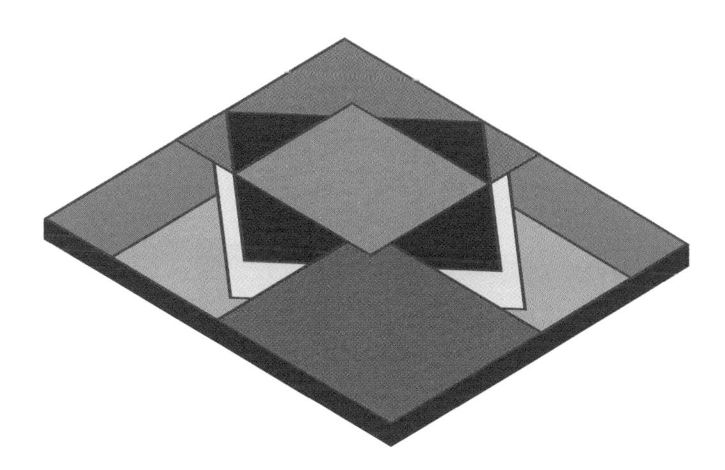

SOLUTION page 132

OVERVIEW

A 2x2x2 cube is made of eight differently coloured cubes of the same size. Four of the cubes have been removed. Five of the six faces are shown, as seen from above. A thicker border indicates that this cube is on the bottom row. When looking at the sixth and final face, as seen from above, what do you see?

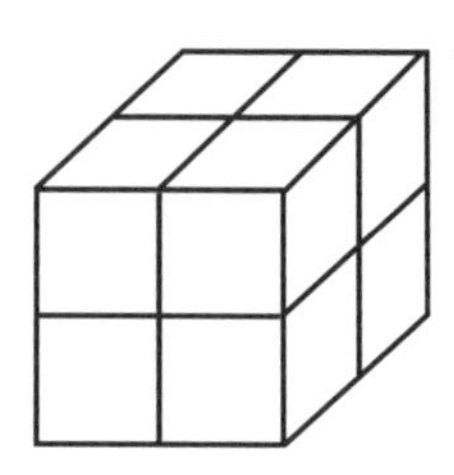

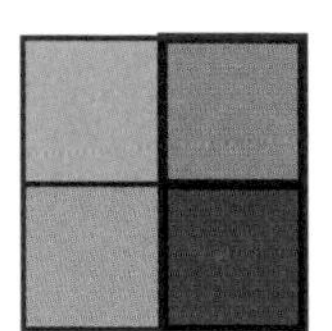

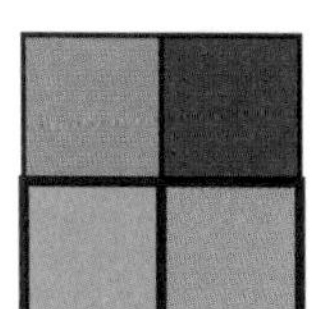

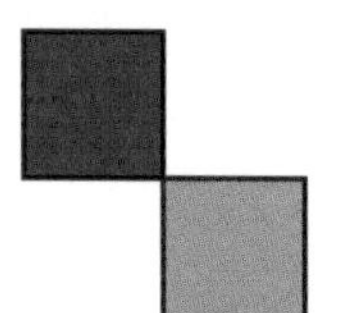

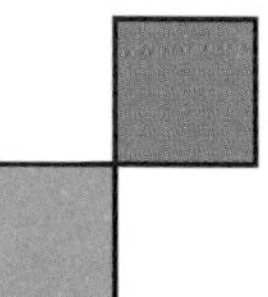

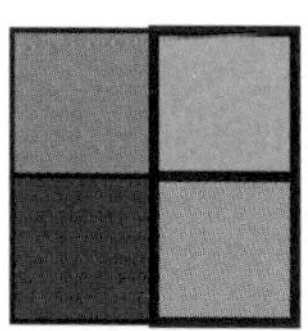

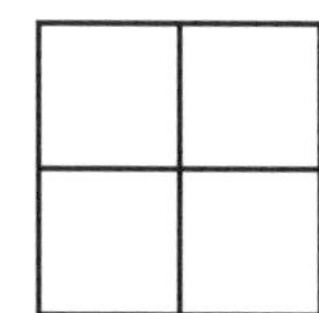

SOLUTION page 139

A 2x2x2 cube is made of eight differently coloured cubes of the same size. Three of the cubes have been removed. Five of the six faces are shown, as seen from above. A thicker border indicates that this cube is on the bottom row. When looking at the sixth and final face, as seen from above, what do you see?

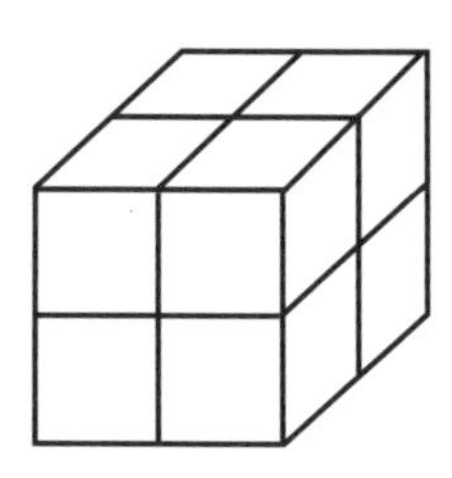

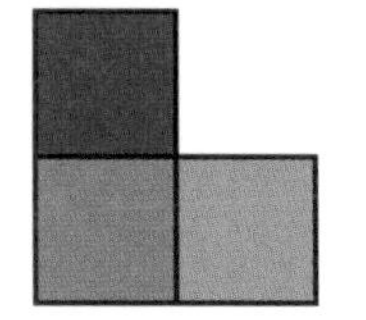

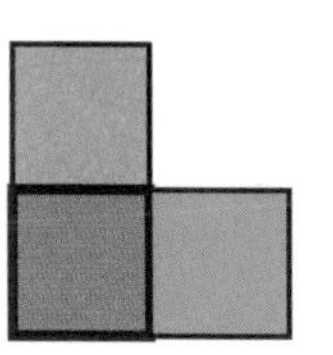

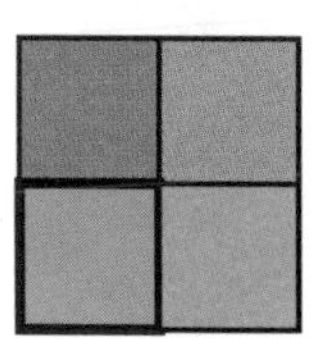

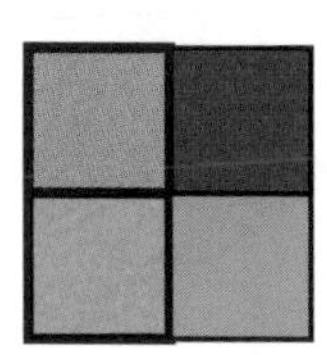

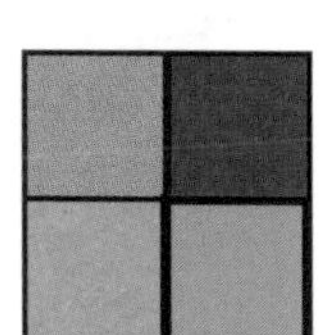

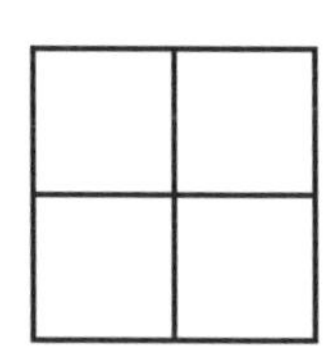

SOLUTION page 132

HOW MANY

How many discrete triangles can be seen in this diagram?

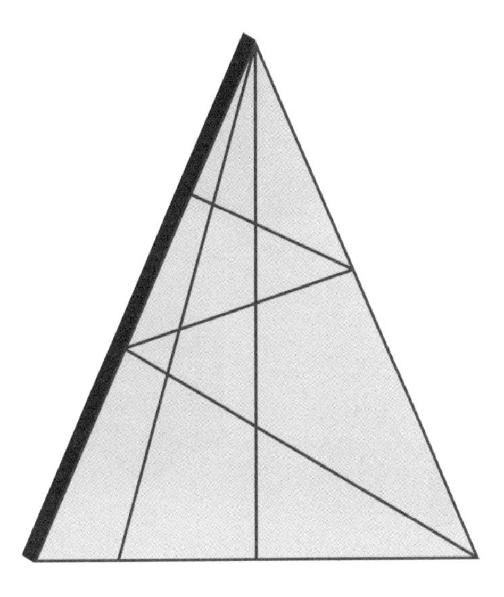

SOLUTION page 139

How many discrete rectangles can be seen in this diagram? You need to work this one out, or you'll be counting for a while...

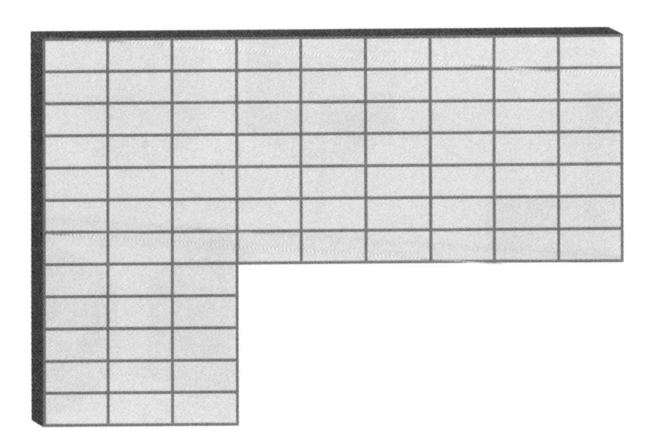

SOLUTION page 139

CUBE NET

181 How many, and which, of the five cubes shown can be formed from the cube net below?

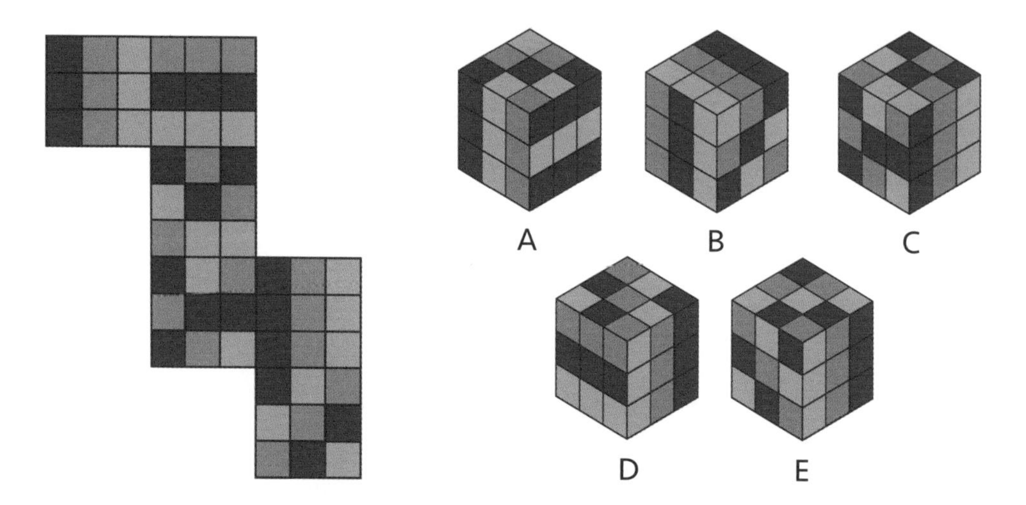

SOLUTION page 133

182 How many, and which, of the five cubes shown can be formed from the cube net below?

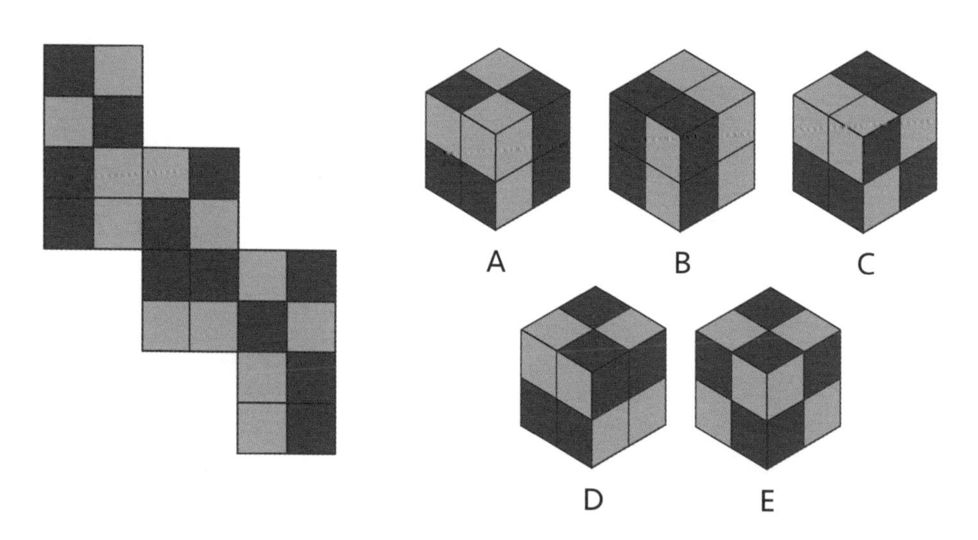

SOLUTION page 140

PERIMETER

This tiled pattern comprises rectangular and square tiles. If the perimeter of a square tile is 48cm, what is the perimeter of the rectangular tiles, and of the overall area?

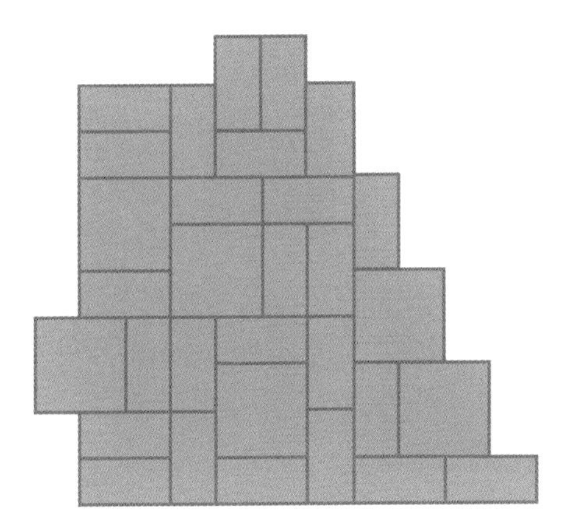

SOLUTION page 140

Each brick in the wall has a perimeter measurement of 50cm. Including holes, what is the overall perimeter of the wall?

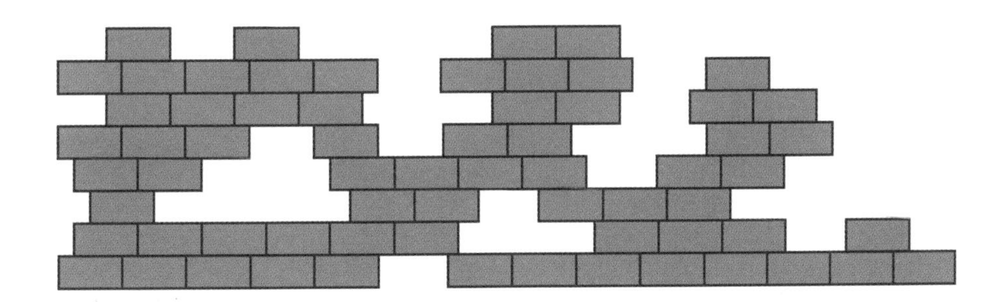

SOLUTION page 140

SHAPE UP

185 If a square sheet of paper is folded along the dotted lines, and cut along the solid lines, which pattern is produced? Cutting is done before folding.

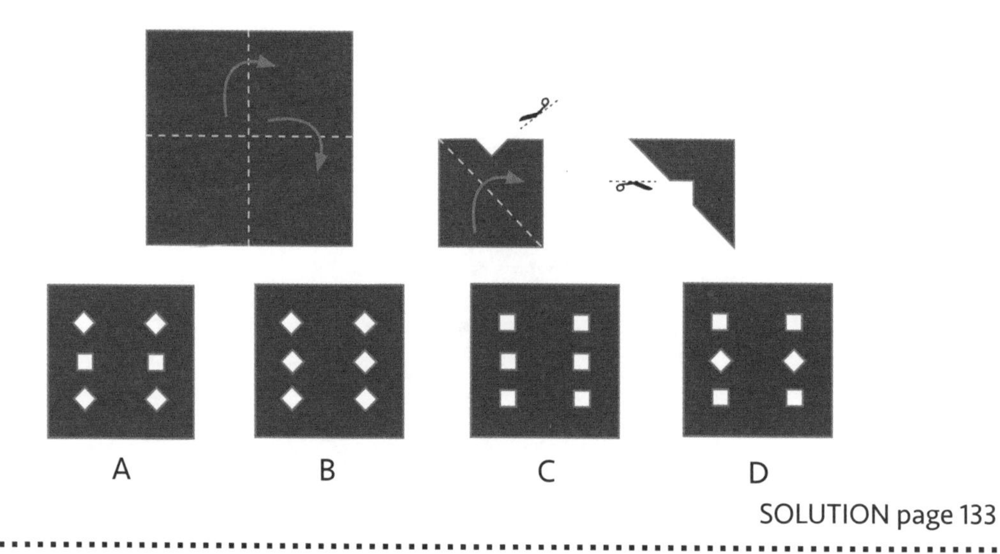

SOLUTION page 133

186 If a square sheet of paper is folded along the dotted lines, and cut along the solid lines, which pattern is produced? Cutting is done before folding.

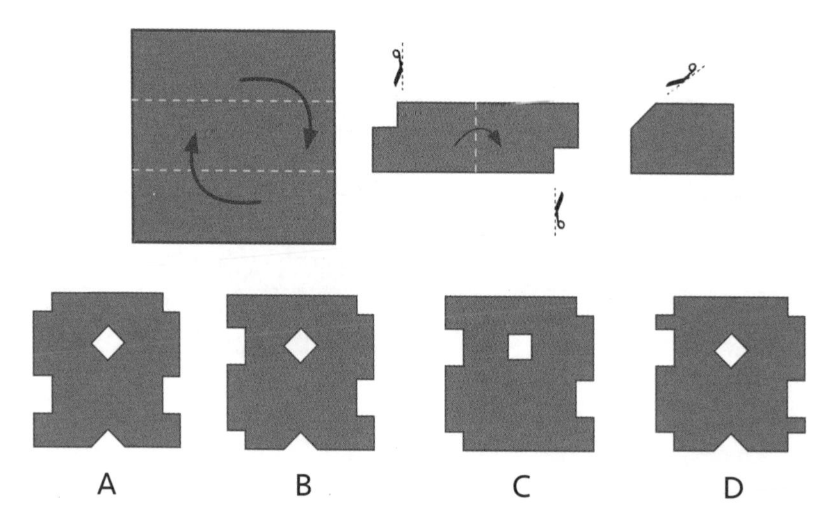

SOLUTION page 140

ANGLE GRINDER

187 Work out the value of the angle marked with a question mark.

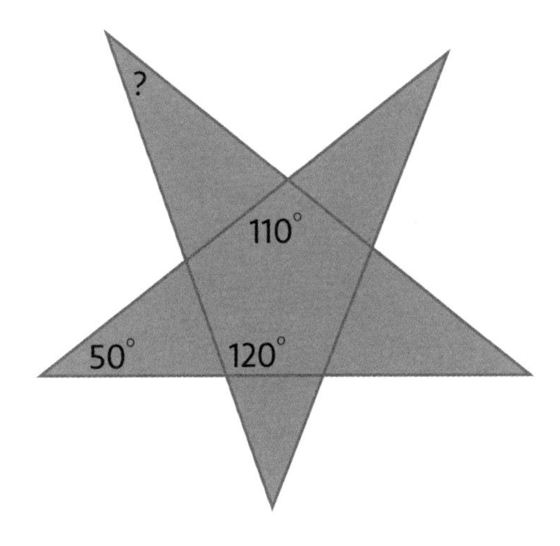

SOLUTION page 140

188 Given that the internal angles of a triangle sum 180°, as do angles on a straight line, what is the measurement of angle H? The star is not drawn to scale.

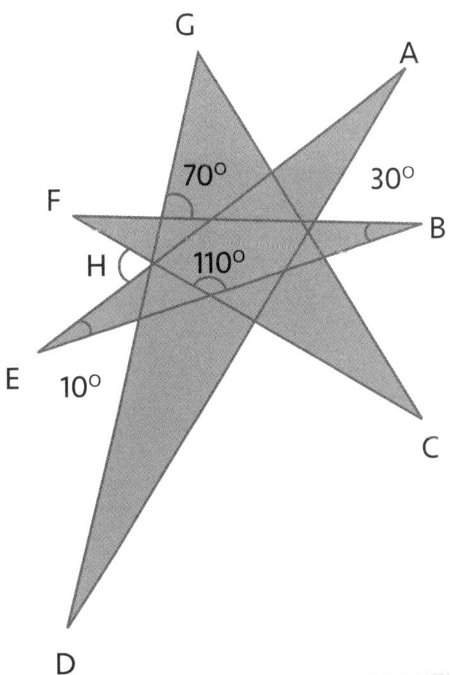

SOLUTION page 133

AREA MAZE (3D)

189 What area is represented by the question mark?

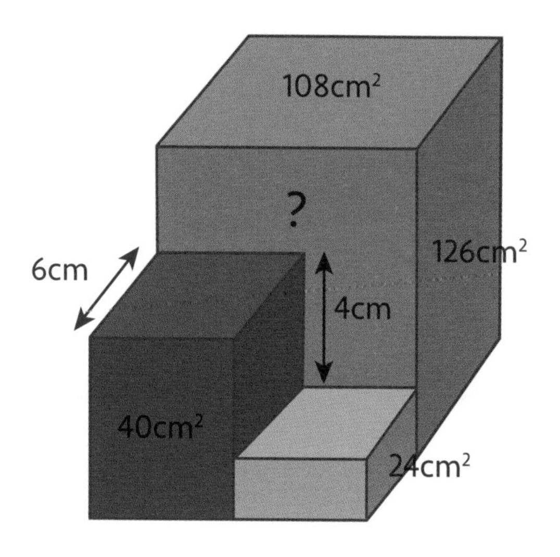

SOLUTION page 134

190 What length is represented by the question mark?

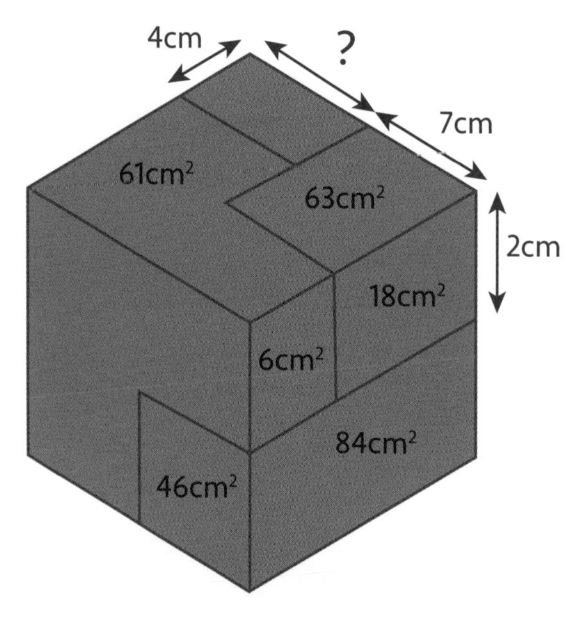

SOLUTION page 141

PENTOMINOES

191 Arrange all twelve pentominoes pictured (as they are and/or rotated and/or reflected) to create the pictured umbrella.

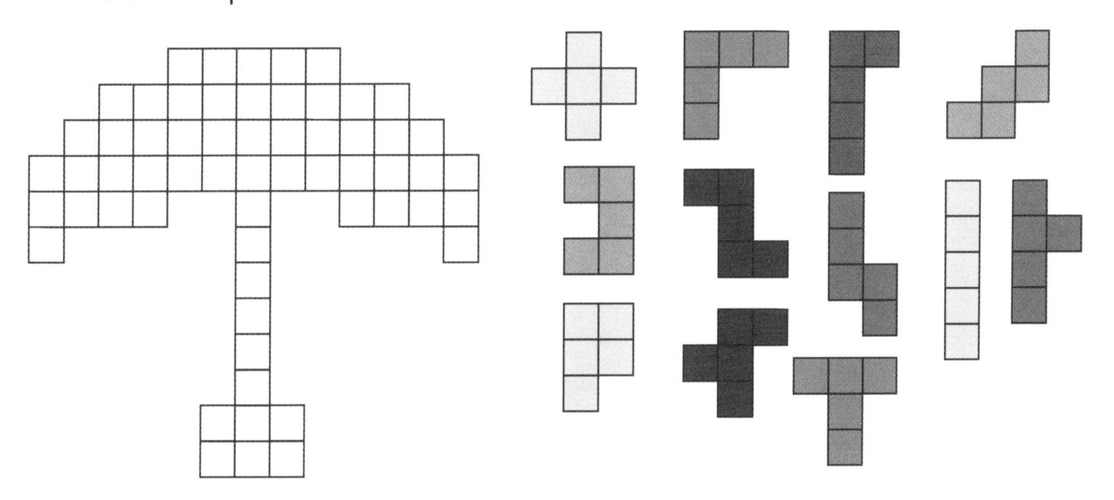

SOLUTION page 135

192 Arrange all twelve pentominoes pictured (as they are and/or rotated and/or reflected) to create the pictured kangaroo.

SOLUTION page 142

193 Coining It In

Twelve discs are arranged to form a square, with four discs on each edge. Using the same twelve discs, arrange them into a square with five discs on each edge.

SOLUTION page 137

194 Coining It In

How can you arrange twelve coins into three horizontal and three vertical rows so that each row contains four coins?

SOLUTION page 142

AREA MAZE

195

What is the area represented by the question mark?

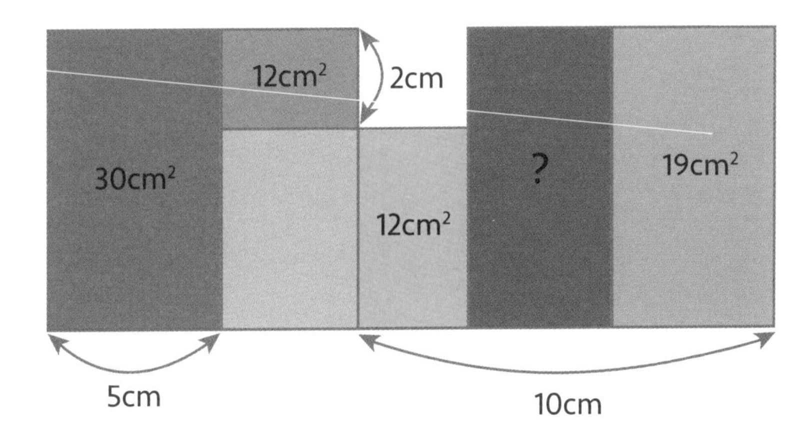

SOLUTION page 129

196

What is the area represented by the question mark?

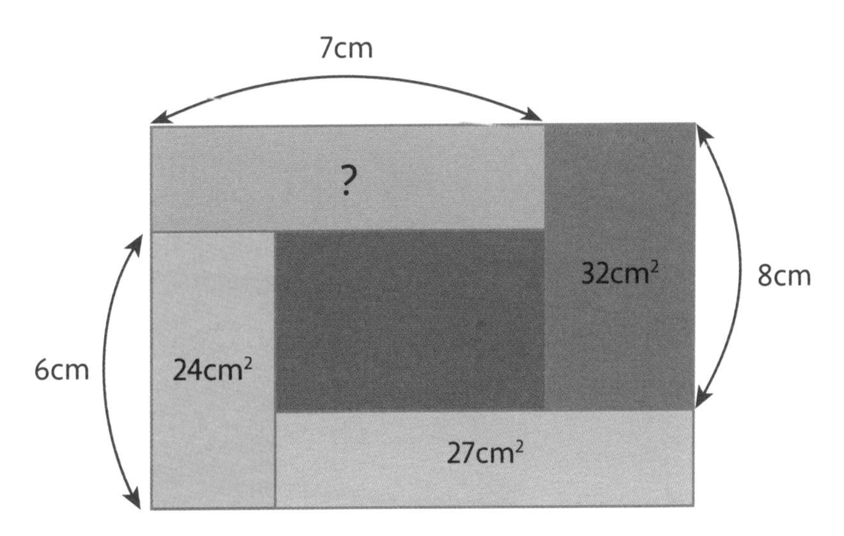

SOLUTION page 142

SPLIT DECISION

197 Split the grid into four identical shapes, following the marked lines. Shapes may be rotated, but not flipped.

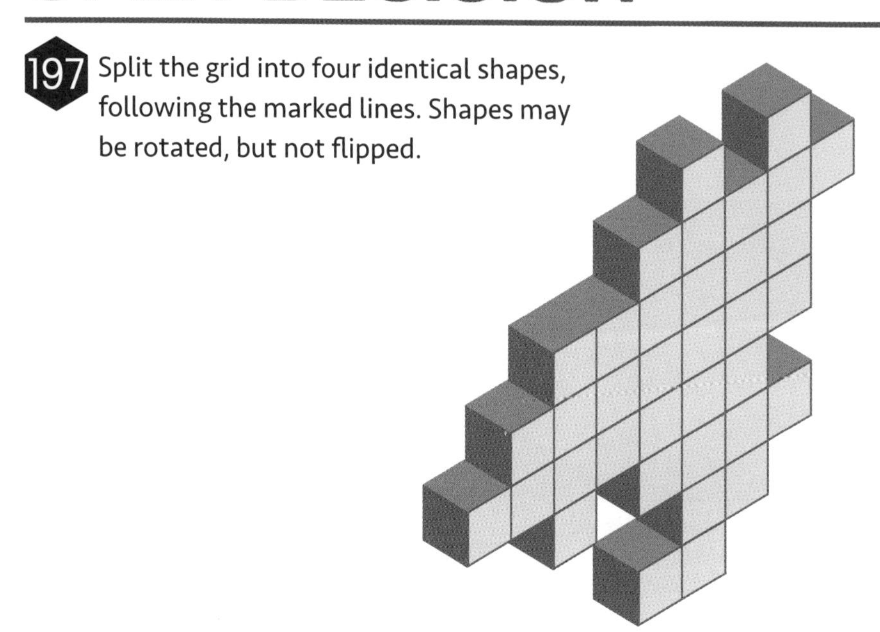

SOLUTION page 143

198 Split the grid into three identical shapes, following the marked lines. Shapes may be rotated, but not flipped.

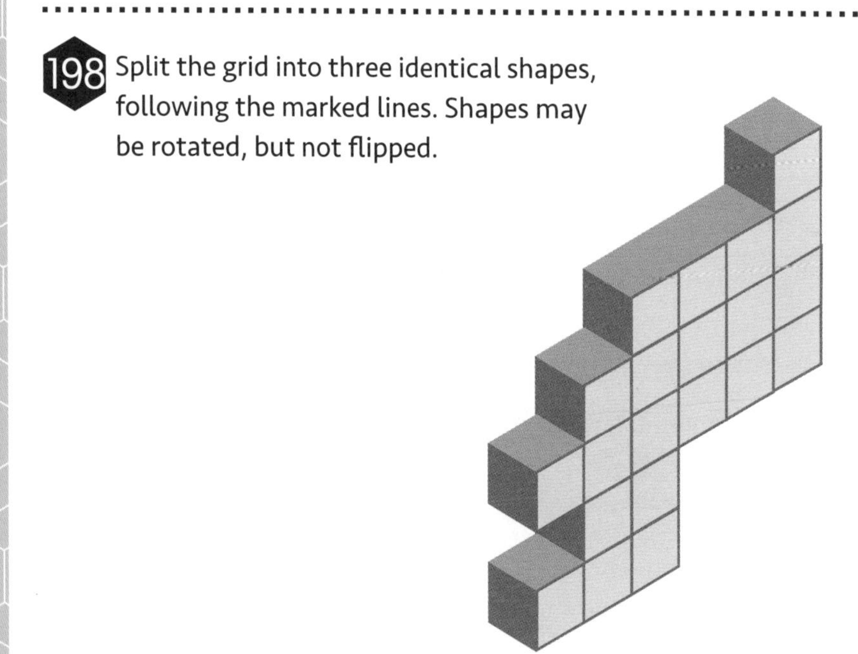

SOLUTION page 143

AREA MAZE (PERCENTAGES)

199 Which area is larger – red or blue?

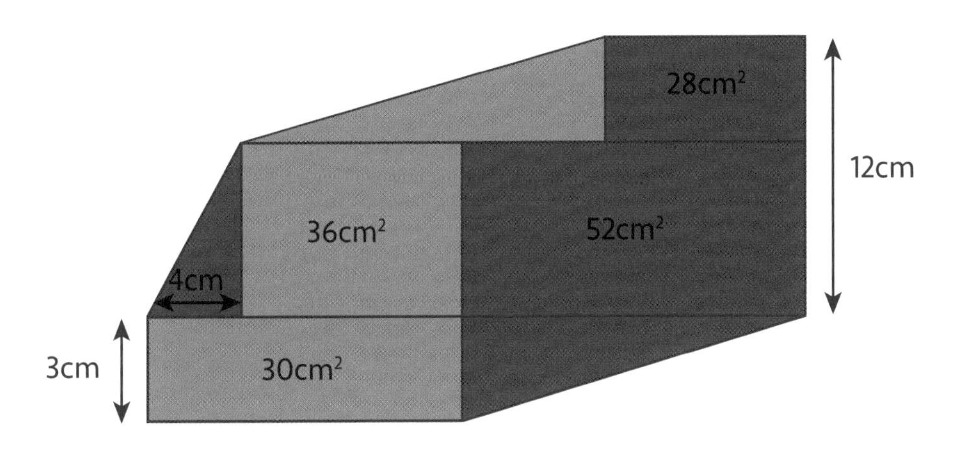

SOLUTION page 143

200 What percentage of the overall shape is represented by the orange area?

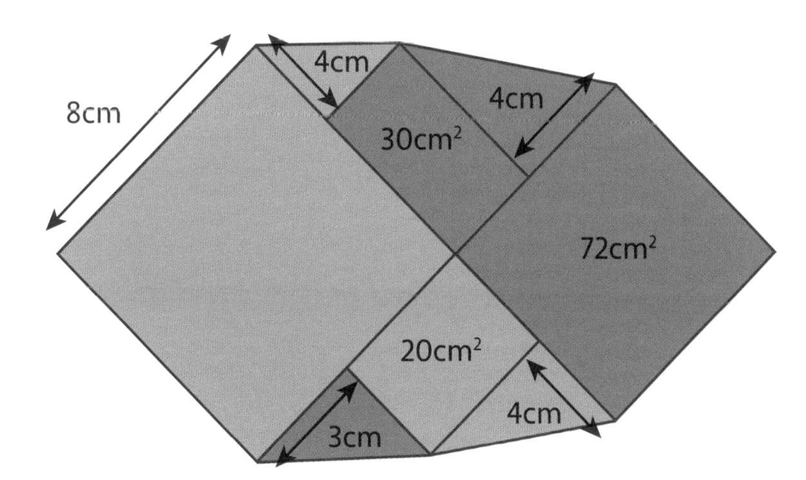

SOLUTION page 136

SOLUTIONS

SOLUTIONS

01

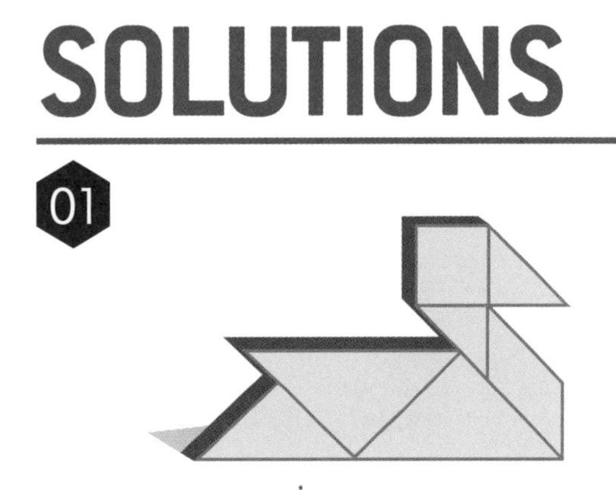

41

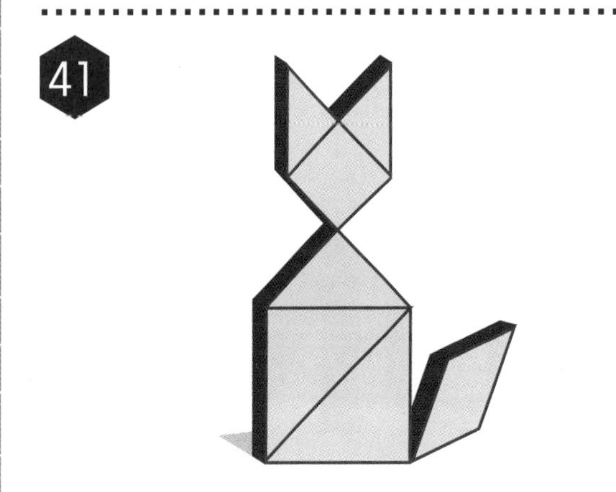

03

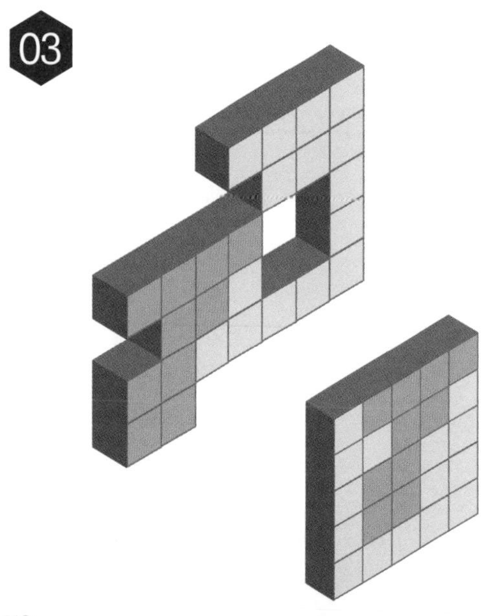

45

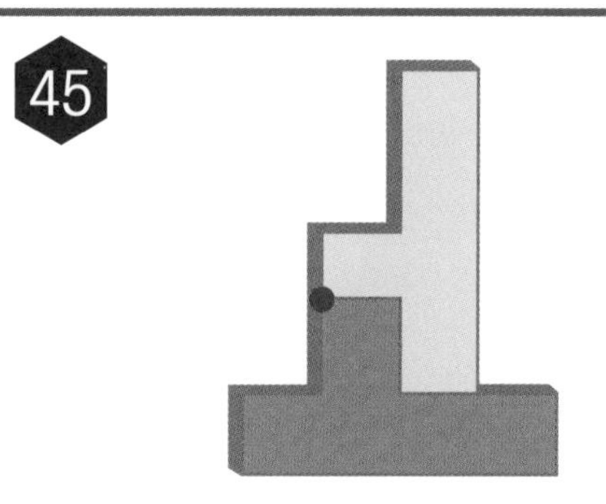

48

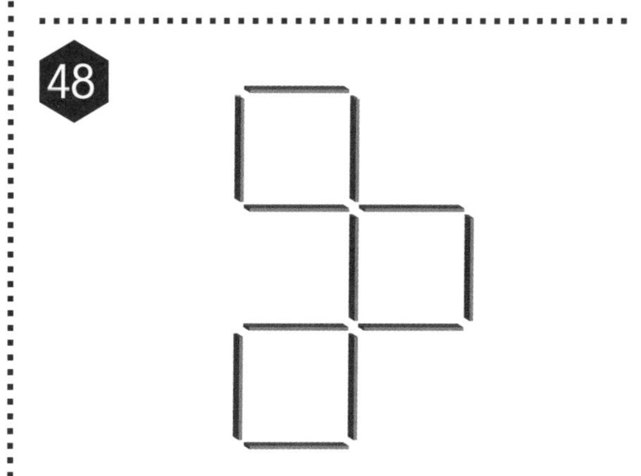

06

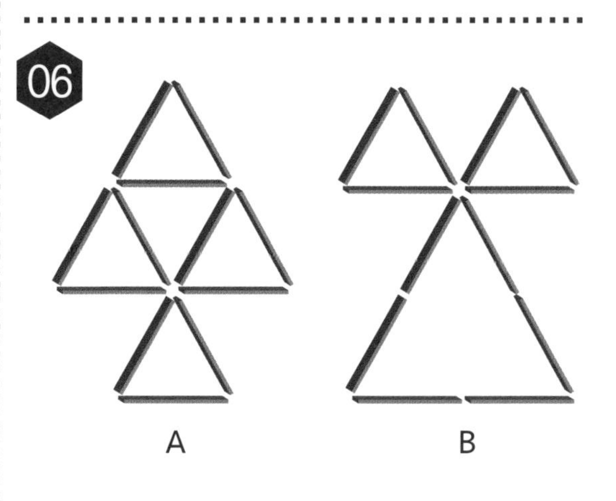

07

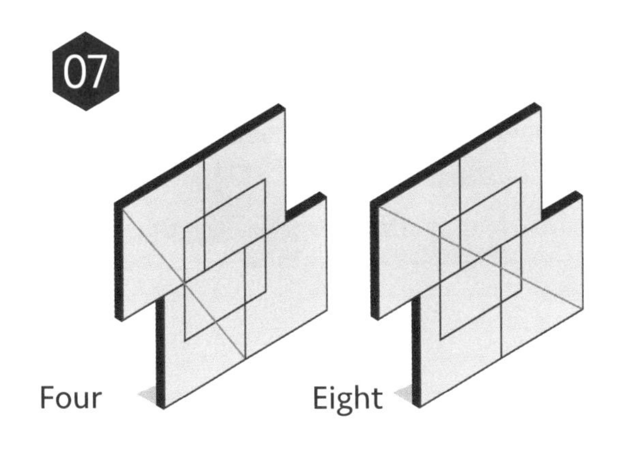

10

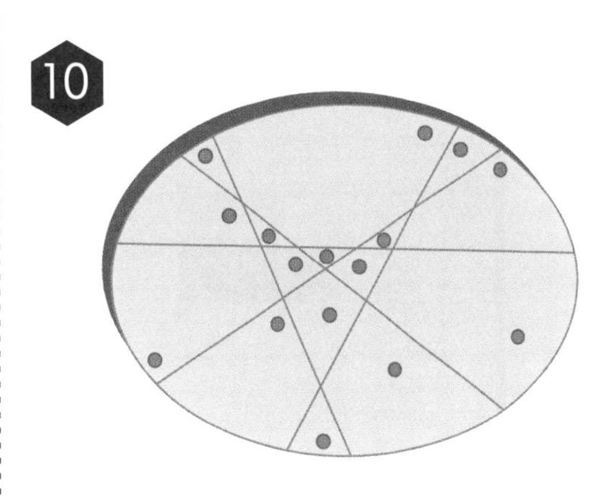

49

20.

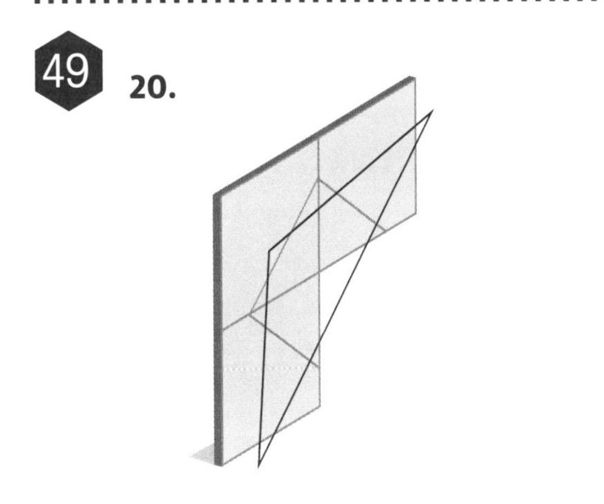

39

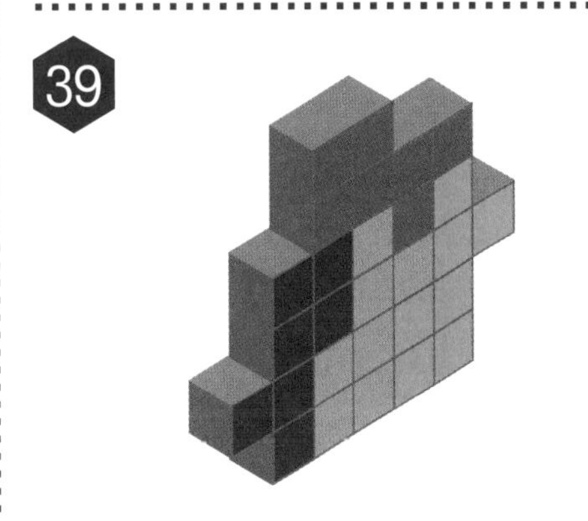

52

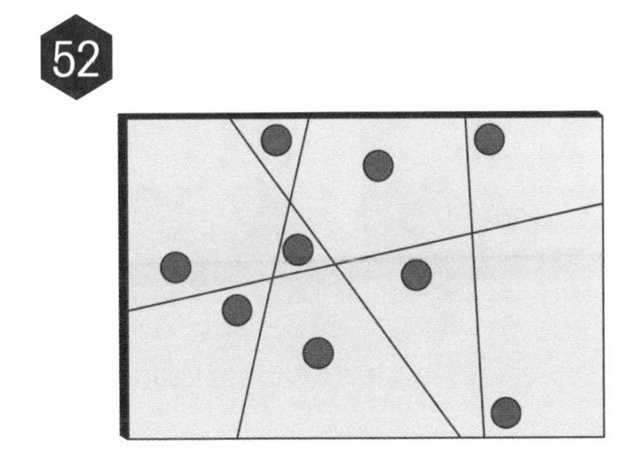

12

SOLUTIONS

55

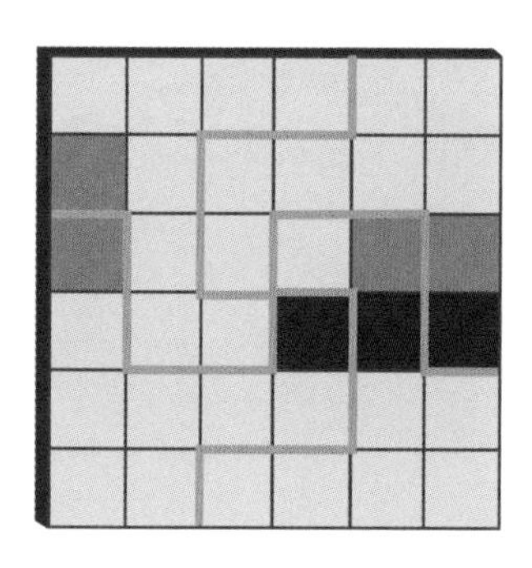

14

15

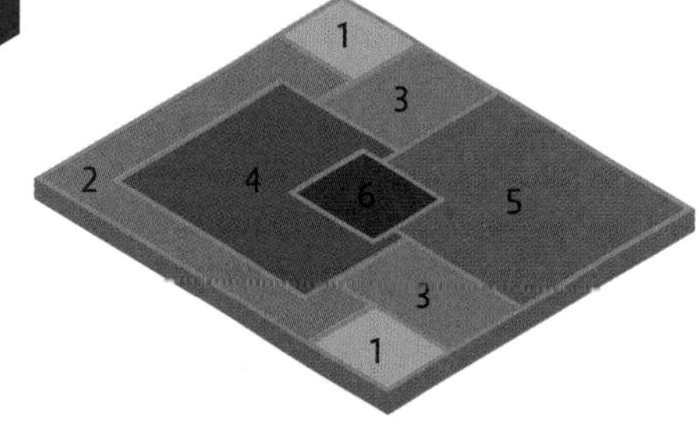

23

180cm.
The edge of the wall comprises just ten lengths and ten heights of brick, equivalent to five bricks.

60

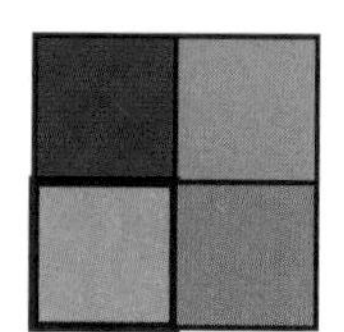

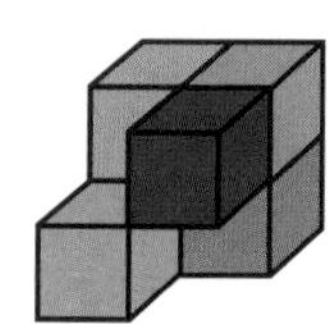

18

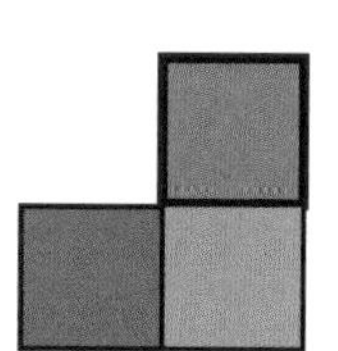

61

16.
10 1x1, 4 2x2, 1 3x3, 1 4x4.

20

45.

64

A

E

24

496cm (13 x 18cm, 19 x 8cm, 22 x 5cm).

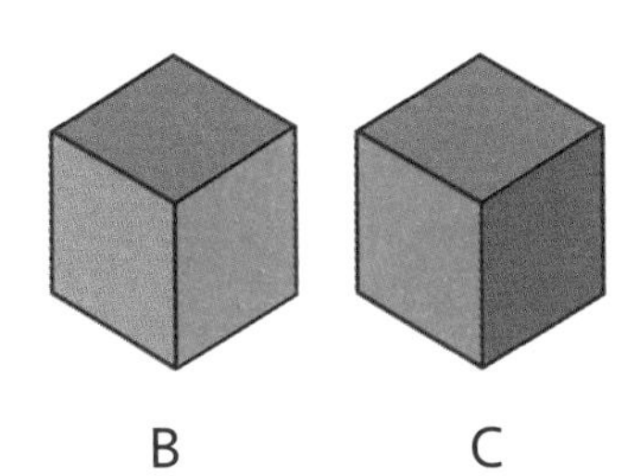

57

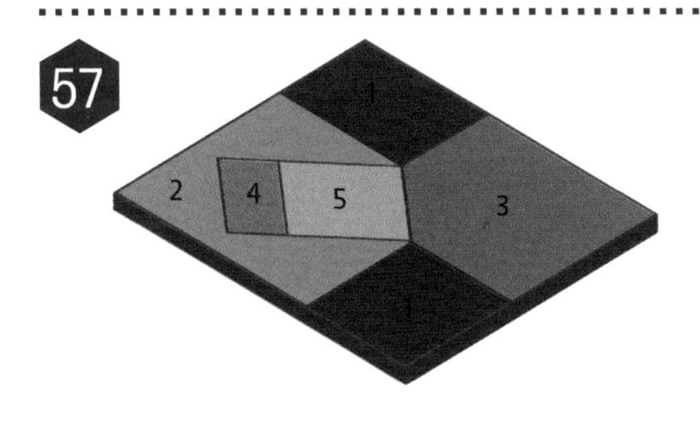

67

A.

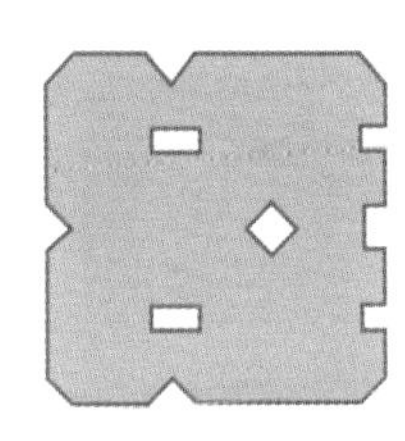

26

A.

27

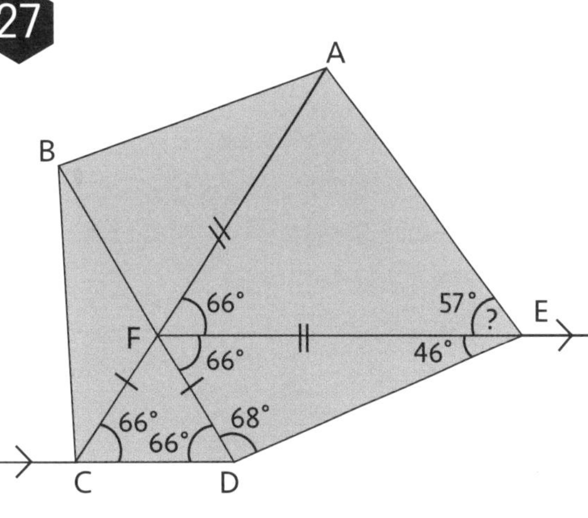

57°.

Angle DFE = $180^{\circ} - 46^{\circ} - 68^{\circ} = 66^{\circ}$

Angle CDF = Angle DFE = 66° (Alternate interior angles)

Angle FCD = Angle CDF = 66° (Isosceles triangle)

Angle AFE = Angle FCD = 66° (Corresponding angles)

Angle FEA =$(180^{\circ} - 66^{\circ})/2 = 57^{\circ}$ (Isosceles triangle)

SOLUTIONS

28

18°.

Each of the internal angles of a polygon is $\{(N-2)\times180^\circ\} \div N$. So, angle X is 108°. Angle Y must be $108^\circ\div2$ as the angle is bisected equally. So angle Y must be 54°, and angle C must be 18°.

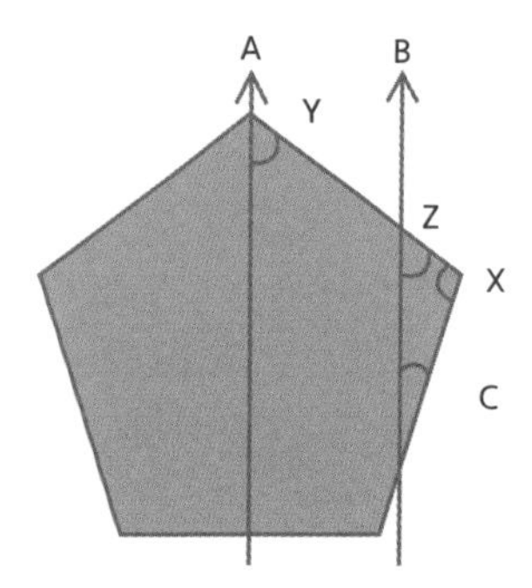

29

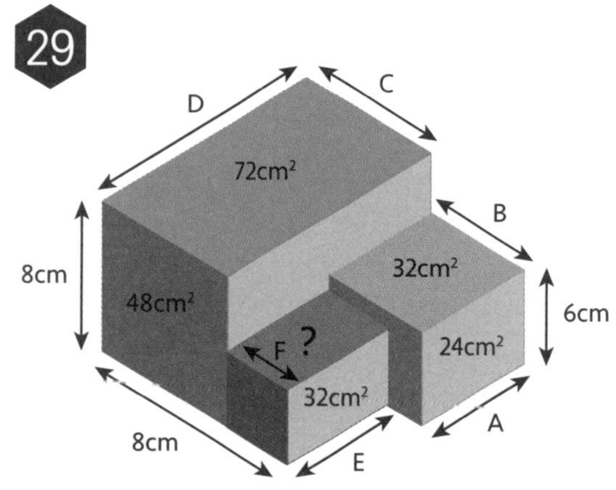

16cm².

A = 24cm² ÷ 6cm = 4cm. B = 32cm² ÷ 4cm = 8cm. C = 48cm² ÷ 8cm = 6cm. D = 72cm² ÷ 6cm = 12cm. E = 12cm – A = 8cm. F = 8cm – C = 2cm.

Therefore the area represented by the questions mark is 16cm².

72

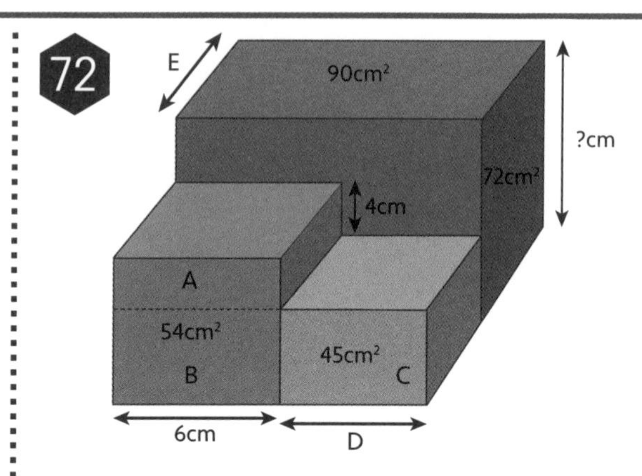

12cm.

Create areas A and B.

A = 6cm x 4cm = 24cm².

B = 54cm² – 24cm² = 30cm².

C = 30cm² ÷ 6cm = 5cm.

D = 45cm² ÷ 5cm = 9cm.

E = 90cm² ÷ 15cm = 6cm.

? = 72cm² ÷ 6cm = 12cm.

73

32

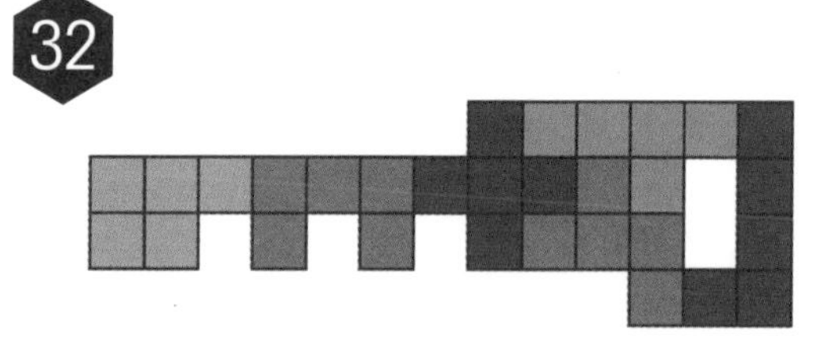

79

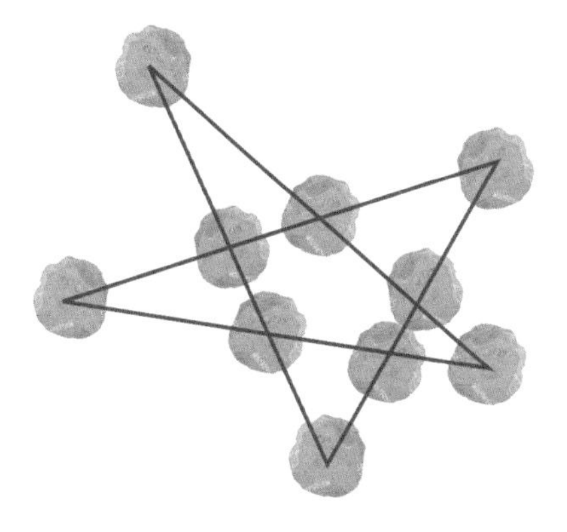

The meterorites are arranged in a star shape.

34

Rather than cut the wood vertically into two congruent pieces, which may be impossible, the only option is to cut the wood in half horizontally, halving the depth, but guaranteeing two identical tabletops.

35

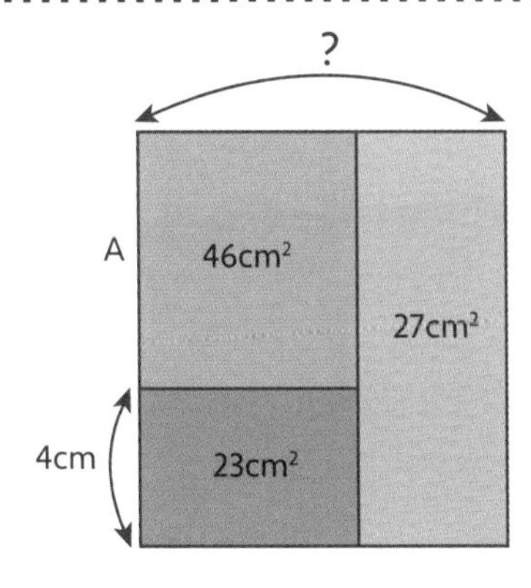

8cm.

As $46cm^2$ is $23cm^2$ x 2, A = 8cm.
The overall area is $46cm^2 + 23cm^2 + 27cm^2 = 96cm^2$.
? = $96cm^2$ ÷ (4cm + 8cm) = 8cm.

78

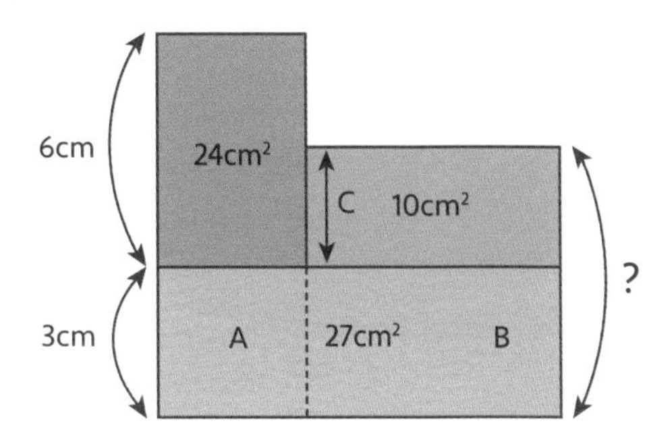

5cm.

Create areas A and B.
A = $24cm^2$ ÷ 2 = $12cm^2$.
B = $27cm^2 - 12cm^2 = 15cm^2$.
$10cm^2 = 15cm^2$ ÷ 3 x 2.
C = 3cm ÷3 x 2 = 2cm.
? = 3cm + 2cm = 5cm.

37

A 54cm² | C 36cm² | J 24cm²
B 216cm² | D 72cm² | H 96cm²
36cm² F | 24cm² E | 16cm² G

130/574, or 23%.

A = B ÷ 4. A is a right-angled triangle, so C = D ÷ = $72cm^2$. E = D ÷ 3, so F (as a right-angled triangle) = B ÷ 6 = $36cm^2$. H = G x 6 = $96cm^2$. As A is to B, so J is to H. Therefore, J = H ÷ 4 = $24cm^2$.

SOLUTIONS

38

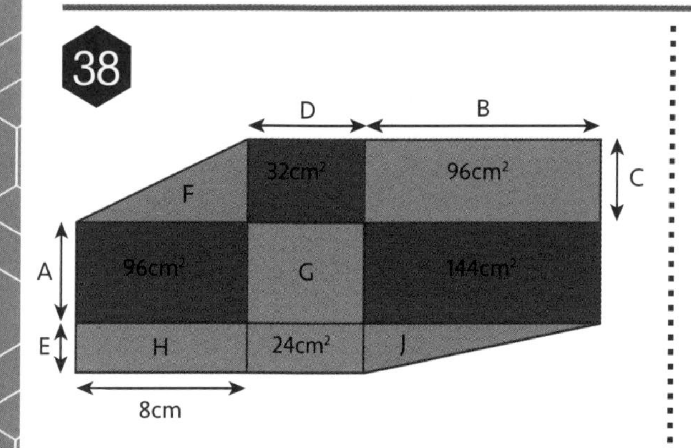

Red: 272; Blue: 284.

A = 96cm^2 ÷ 8cm = 12cm.

B = 144cm^2 ÷ 12cm = 12cm.

C = 96cm^2 ÷ 12cm = 8cm.

D= 32cm^2 ÷ 8cm = 4cm.

E = 24cm^2 ÷ 4cm = 6cm.

F = (C x 8cm) ÷ 2 = 32cm^2.

G = A x D = 48cm.

H = E x 8cm = 48cm^2.

J = (B x E) ÷ 2 = 36cm^2.

11

40

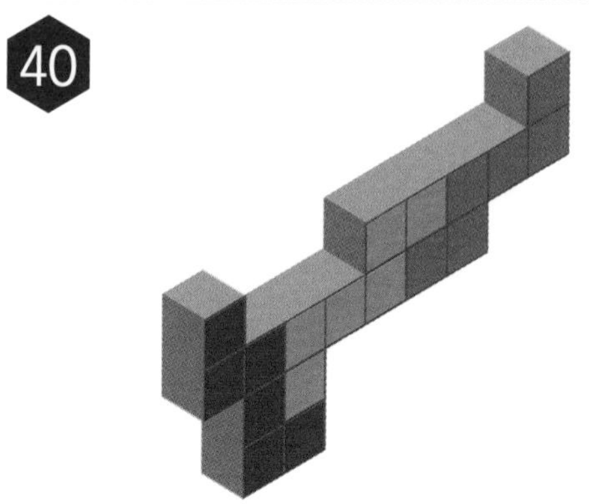

02

42

119

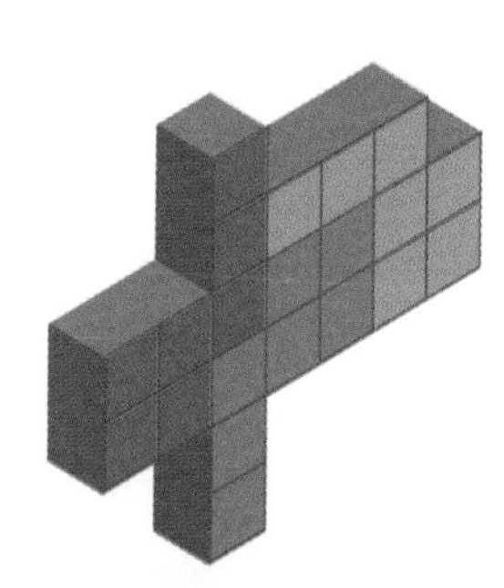

44

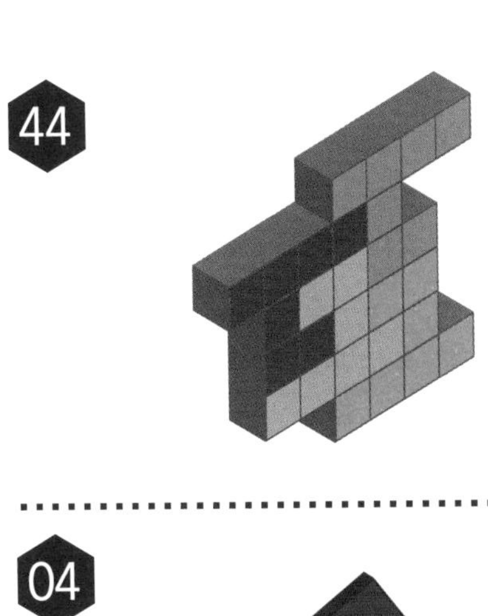

04

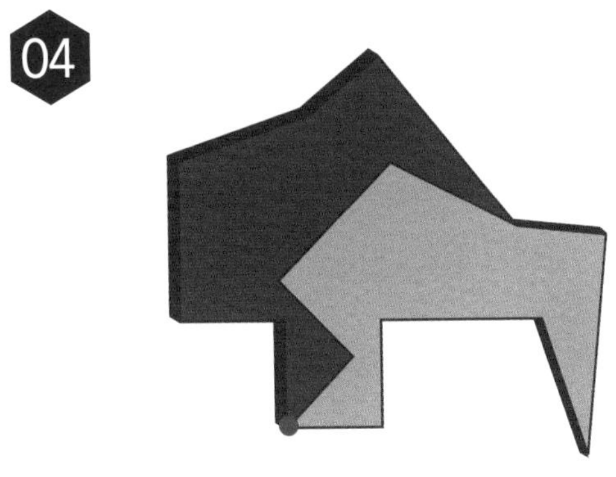

46

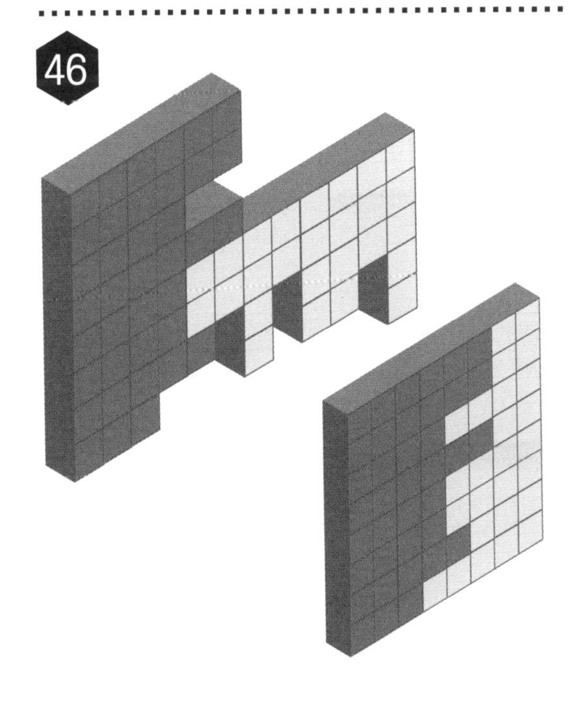

47

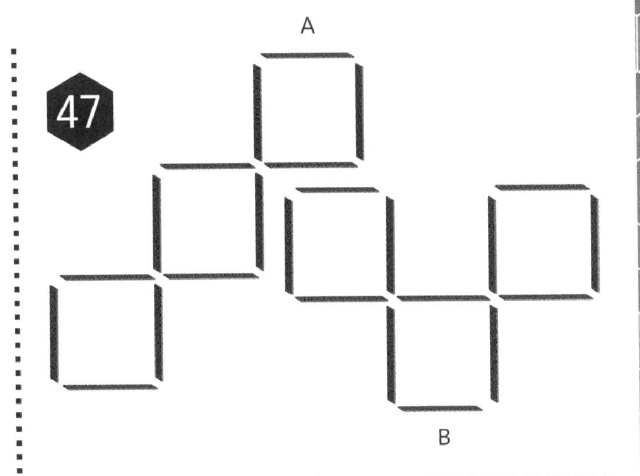

05

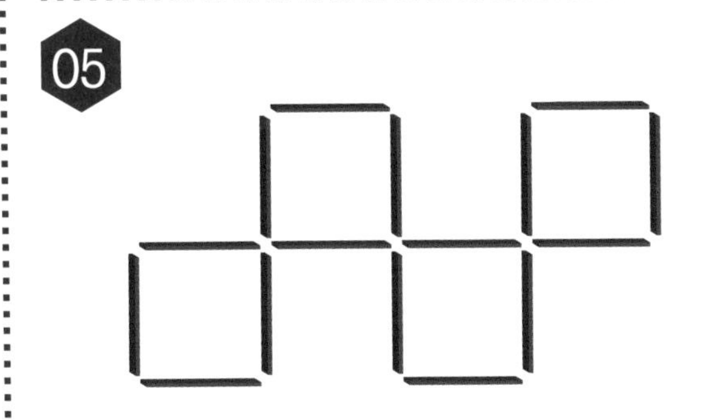

08

SOLUTIONS

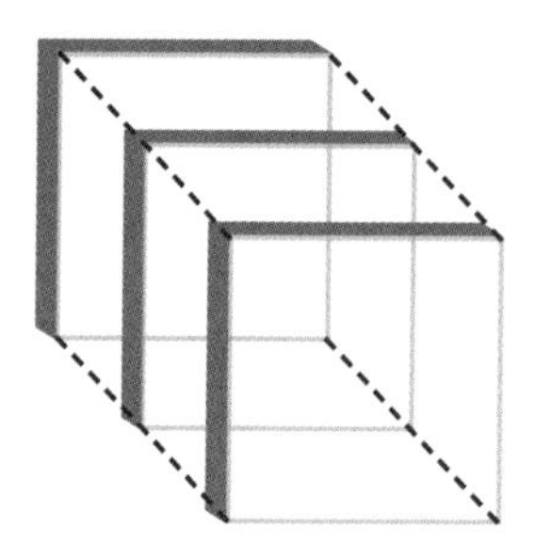

The four triangles are 2D,
the eight squares 3D.

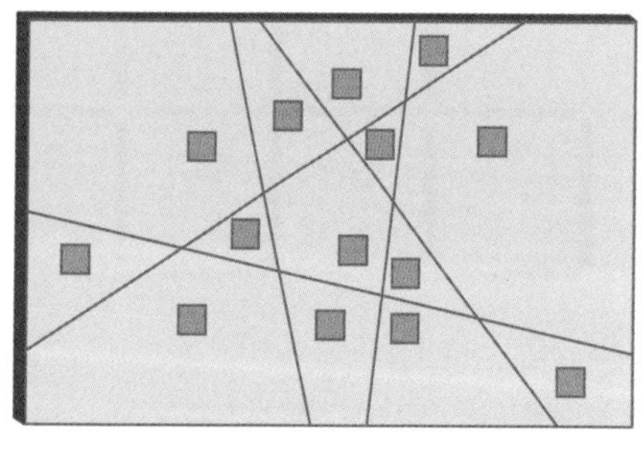

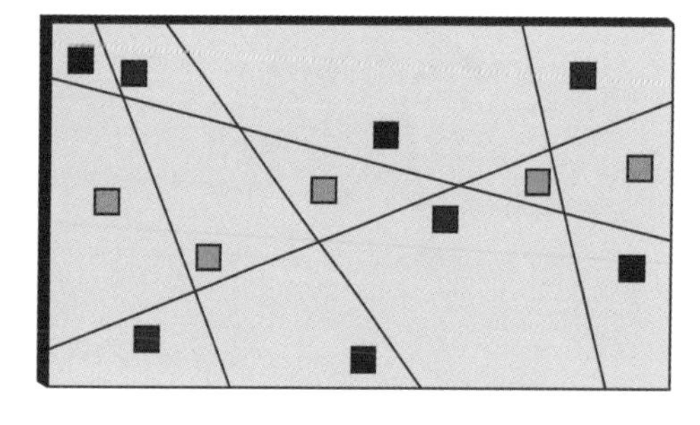

53

54

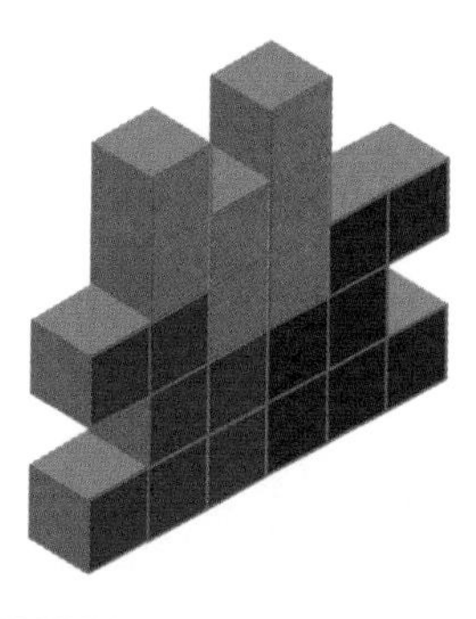

13

91

94

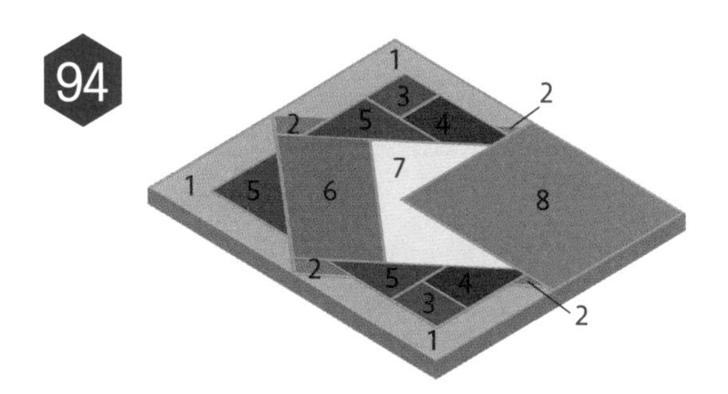

58

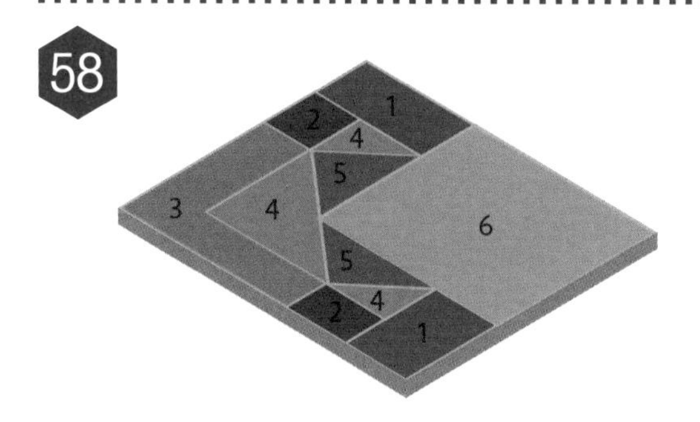

59

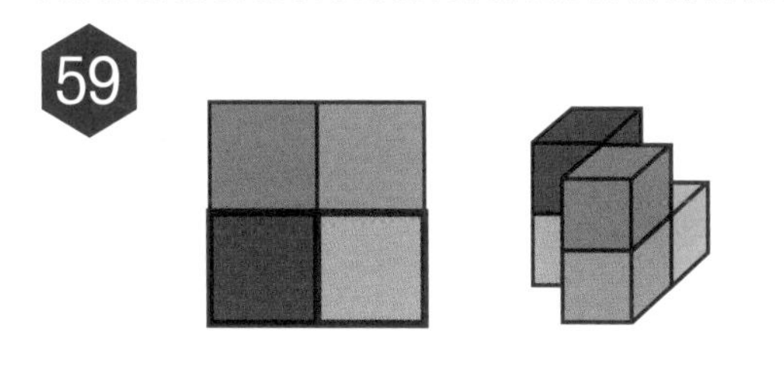

17

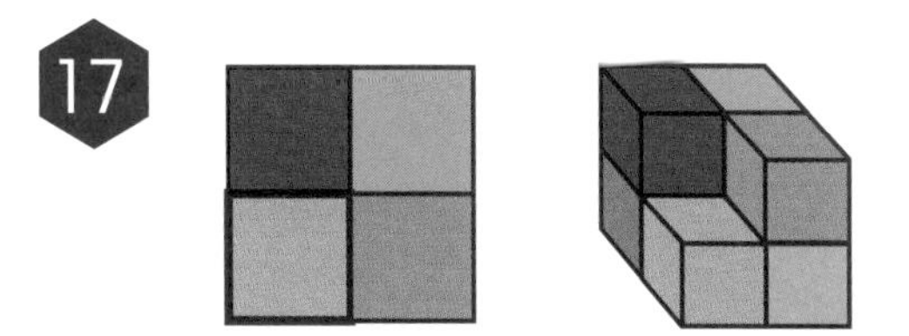

19

35.

62

124.

40 (x1), 28 (x2), 24 (x4), 12 (x8),
8 (x9), 8 (x16), 4 (x18).

63

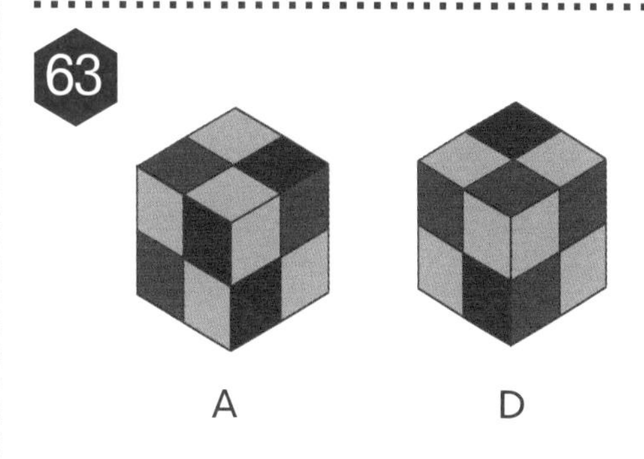

21

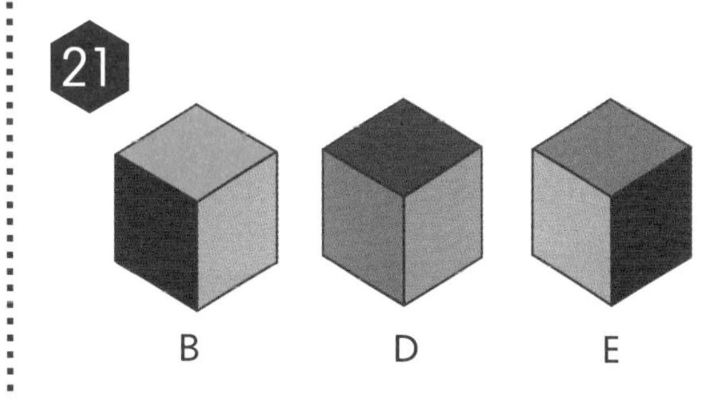

65

Each external edge of the diamond tile is 28cm ÷ 4 = 7cm. This must be the same for the edge of each tile. The yellow area has a perimeter of 16 x 7cm = 122cm. The overall perimeter comprises 32 edges, so the perimeter of the yellow area is half the perimeter of the overall area.

SOLUTIONS

6cm.

The overall perimeter is equivalent to 84 widths, so each block is 504cm ÷ 84 = 6cm.

D.

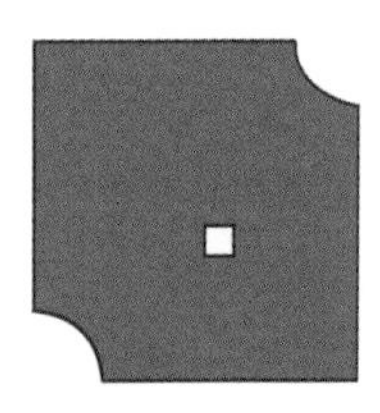

C.

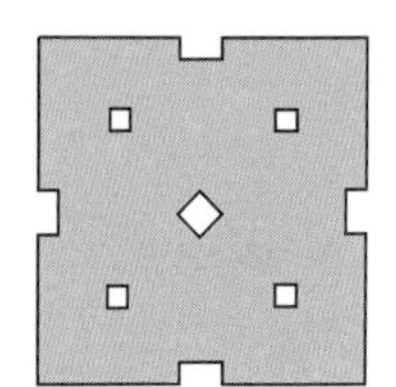

50°.

Angle A = Angle B (opposite angles).

So, 5x – 20° = 4x + 10°.

Therefore, 5x = 4x +10° + 20°, and 5x – 4x = 10° + 20°.

So, x = 30°, and angles A and B = 130°.

Therefore, angle C = 50° (angles on a straight line always total 180°).

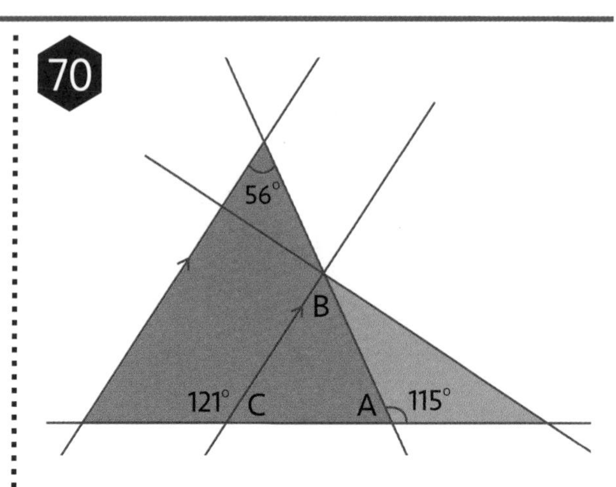

121°.

For ease of reference, add A, B and C to the diagram. Angle A = 180° – 115° = 65°. Angle B = 56° (corresponding angles - straight line transversing parallel lines). Angle C = 180° - 65° - 56° = 59°

Angle ? = 180° - 59° = 121°

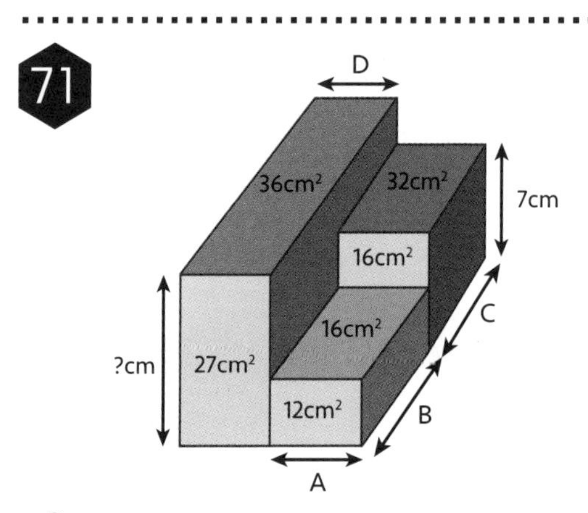

9cm.

A = (16cm² + 12cm²) ÷ 7cm = 4cm.

B = 16cm² ÷ 4cm = 4cm.

C = 32cm² ÷ 4cm = 8cm.

D = 36cm² ÷ 12cm = 3cm.

? = 27cm² ÷ 3cm = 9cm.

30

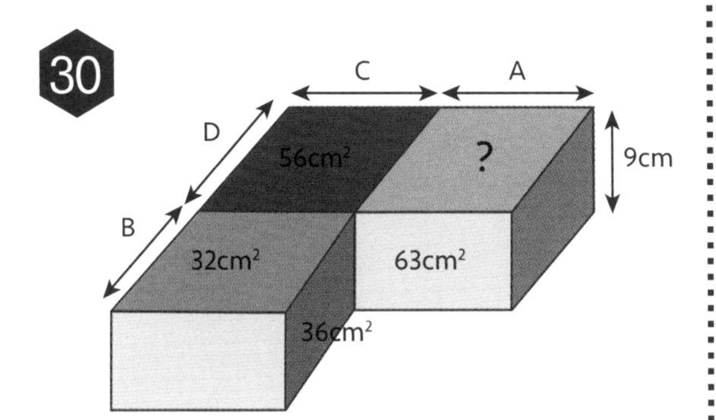

$49cm^2$.

$A = 63cm^2 \div 9cm = 7cm$.

$B = 36cm^2 \div 9cm = 4cm$.

$C = 32cm^2 \div 4cm = 8cm$.

$D = 56cm^2 \div 8cm = 7cm$.

$? = A \times D = 7cm \times 7cm = 49cm^2$.

31

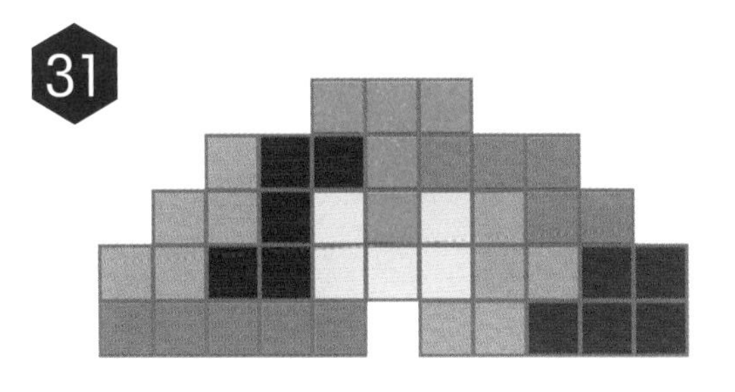

74

75

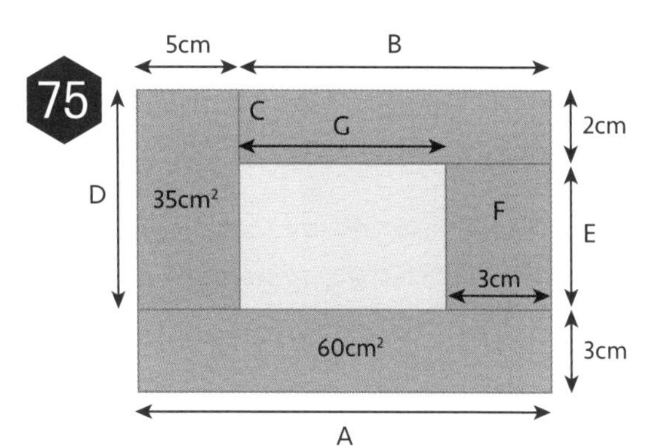

30%.

$A = 60cm^2 \div 3cm = 20cm$.

$B = 20cm - 5cm = 15cm$.

$C = 15cm \times 2cm = 30cm^2$.

$D= 35cm^2 \div 5cm = 7cm$.

$E = 7cm - 2cm = 5cm$.

$F = 5cm \times 3cm = 15cm^2$.

$G = 20cm - 5cm - 3cm = 12cm$.

$H = E \times G = 60cm^2$.

$60cm^2 = 30\%$ of $200cm^2$.

117

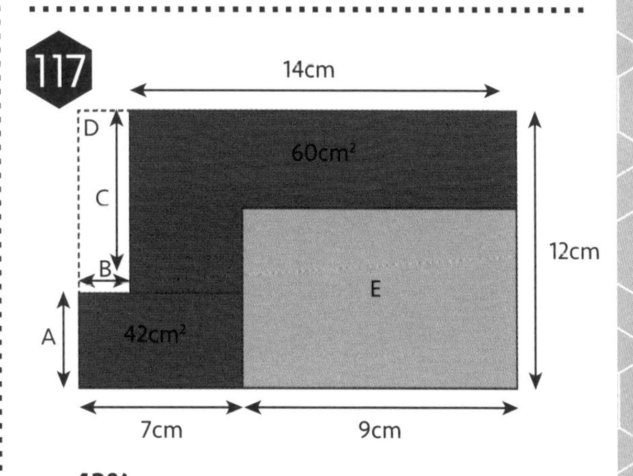

43%.

$A = 42cm^2 \div 7cm = 6cm$.

$B = 9cm + 7cm - 14cm = 2cm$.

$C = 12cm - A = 6cm$.

Area $D = B \times C = 12cm$.

$E = 12cm \times (7cm + 9cm) - D - 60cm^2 - 42cm^2 = 192cm^2 - 114cm^2 = 78cm^2$.

SOLUTIONS

77

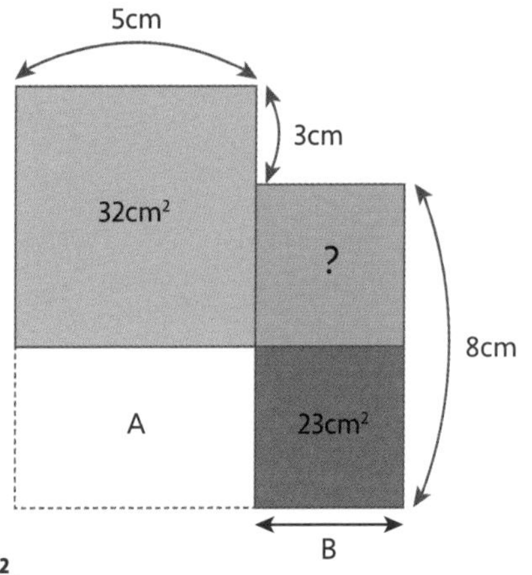

17cm².

Create area A.

A = ((8cm + 3cm) x 5cm) - 32cm² = 23cm².

A (23cm²) = 23cm², so B = 5cm.

? = 5cm x 8cm – 23cm² = 17cm².

36

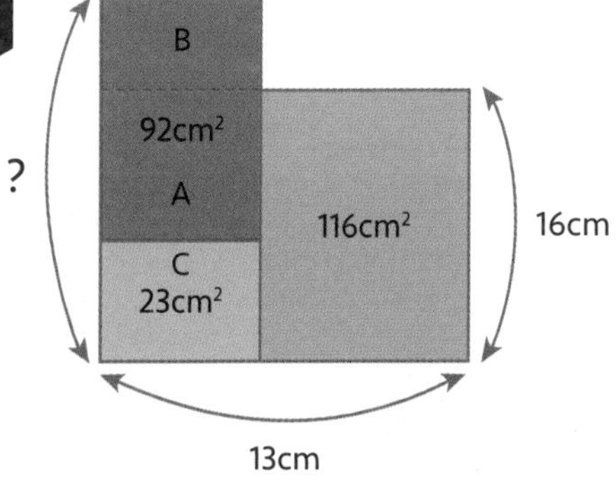

20cm.

Create areas A and B.

A = 13cm x 16cm – 116cm² – 23cm² = 69cm².

B = 92cm² – 69cm² = 23cm².

23cm² = 69cm² ÷ 3.

C = 16cm ÷ 4 = 4cm.

? = 16cm + 4cm = 20cm.

33

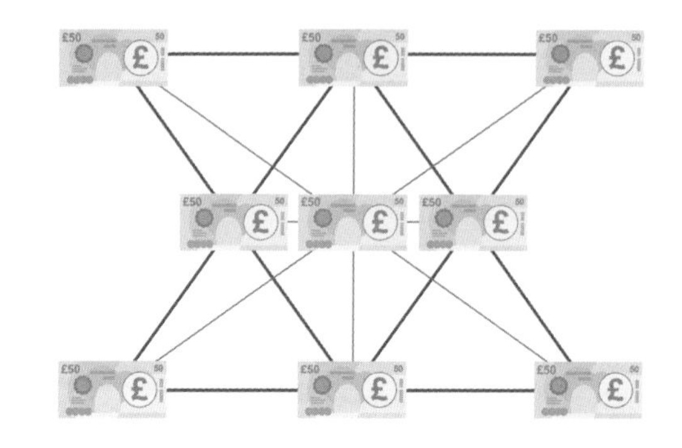

80

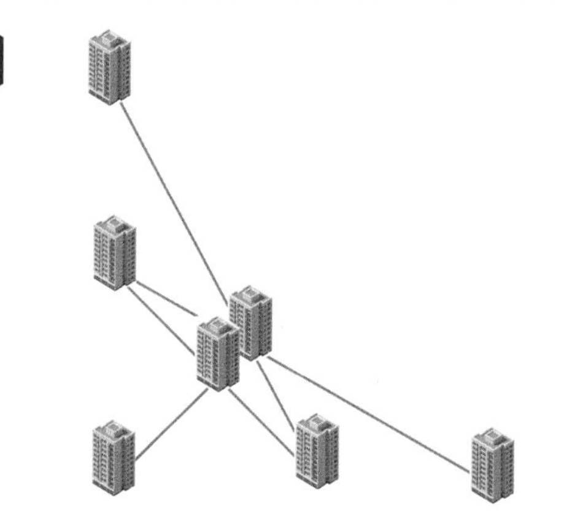

122

82

85

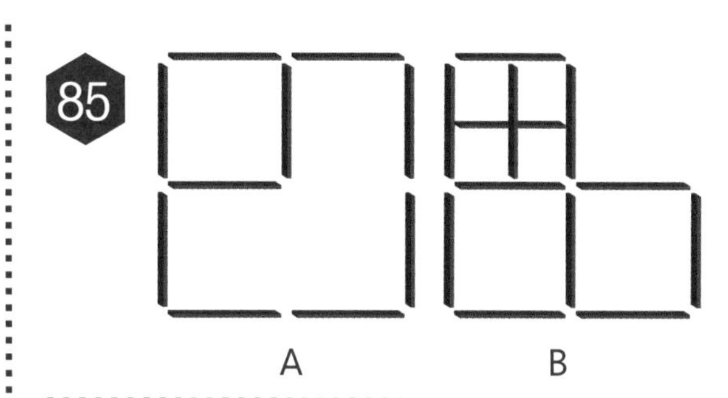

114

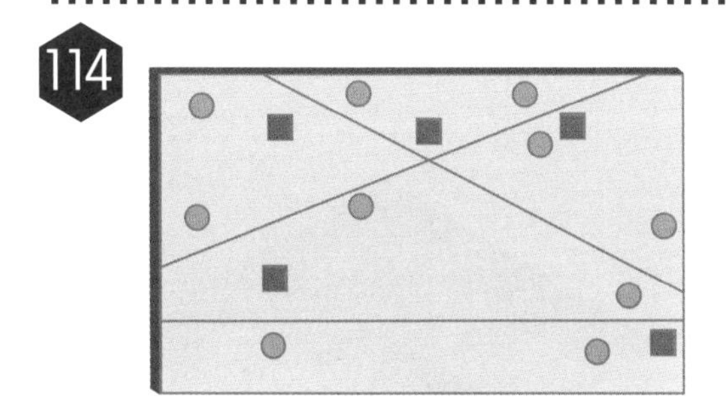

125

84

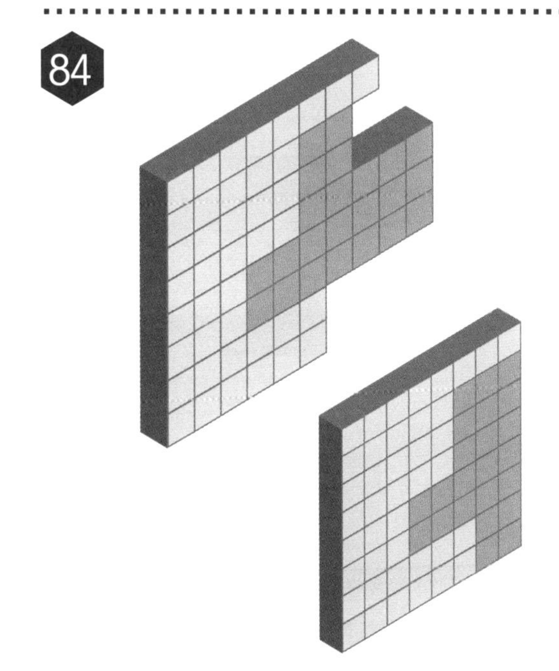

87 **28**

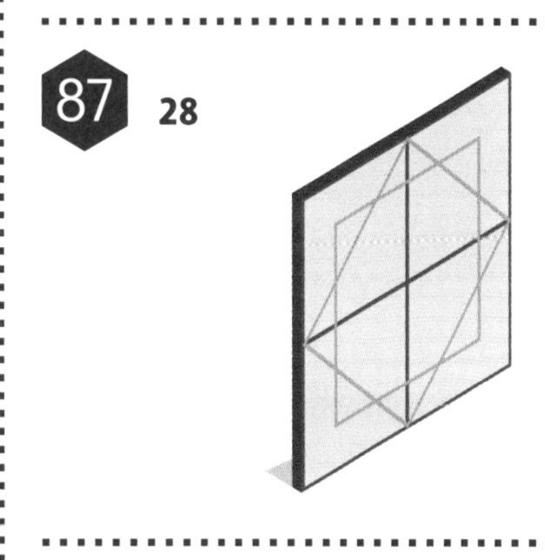

88

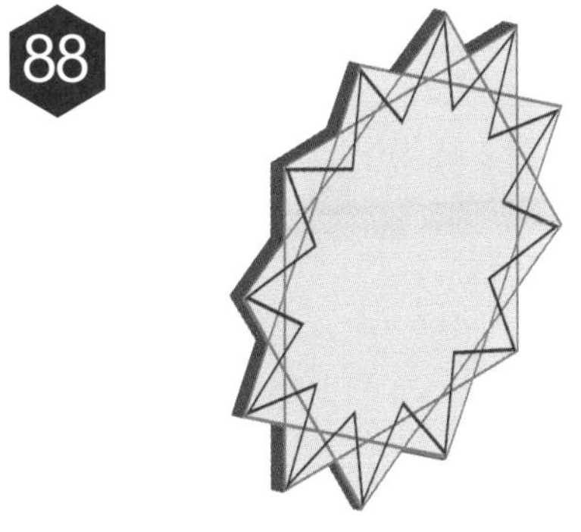

A twelve-pointed star.

SOLUTIONS

89

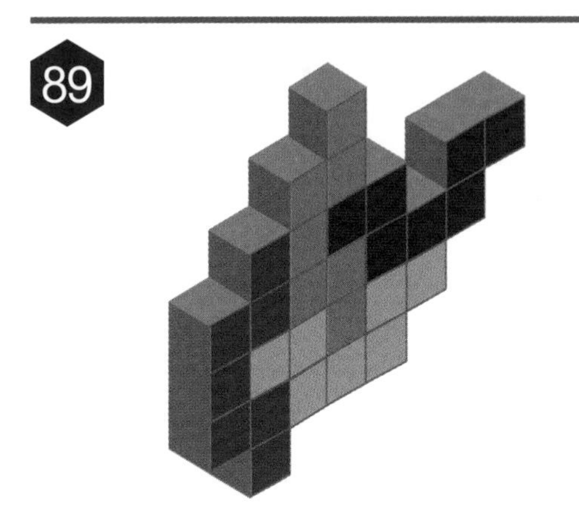

16

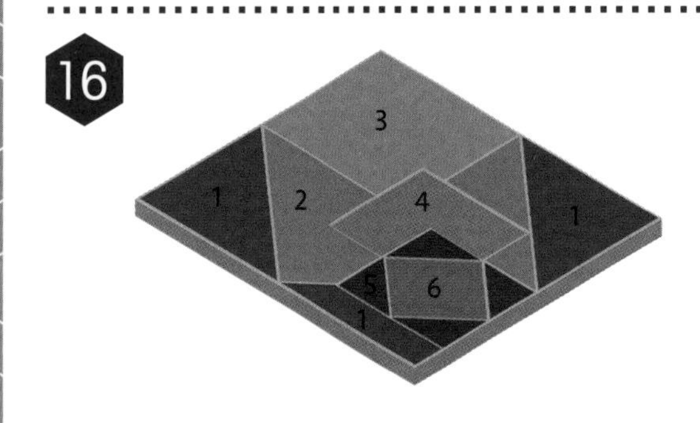

133

102

128cm.

92

93

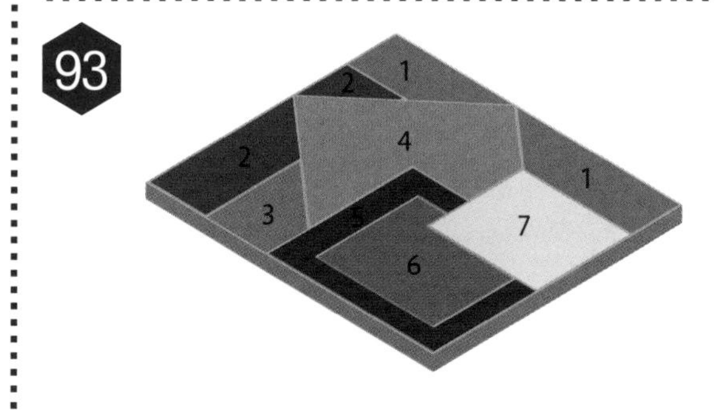

104 **B.**

95

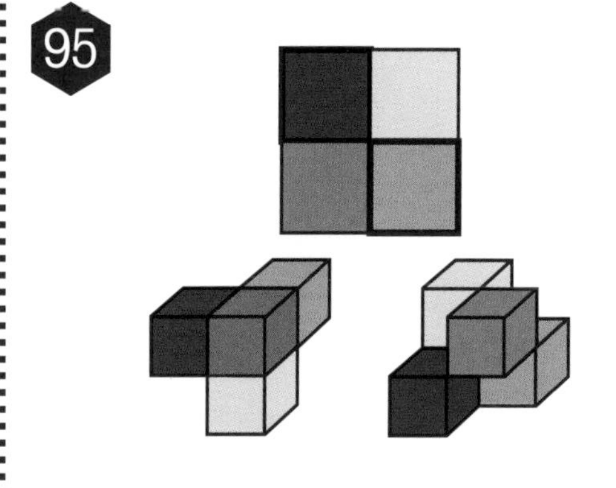

96

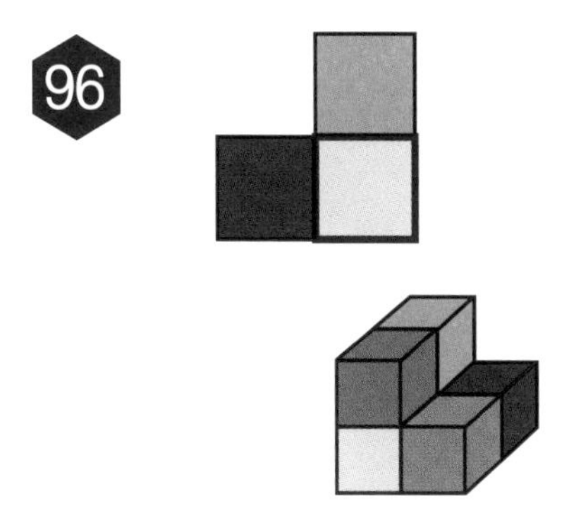

204.

There are 204 squares in total:
8x8 1x1 squares, 7x7 2x2 squares etc.,
so the total is 64+49+...+1=204.

The mathematical formula is N(N+1)(2N+1)÷6, where N is the number of squares in a single row/column.

83

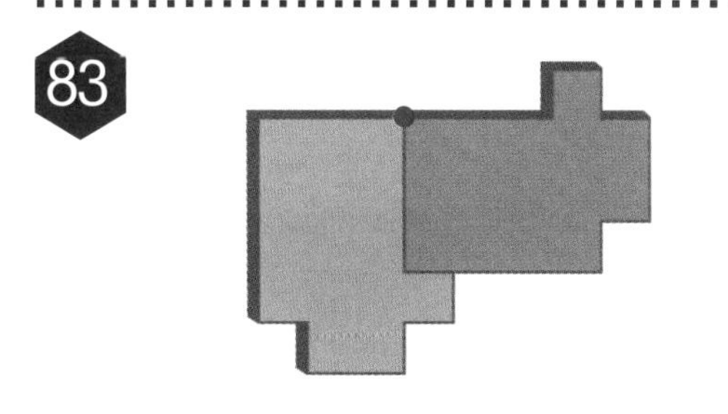

141

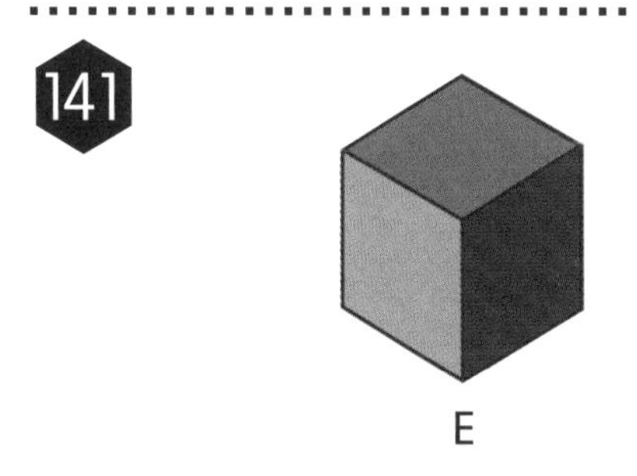

99

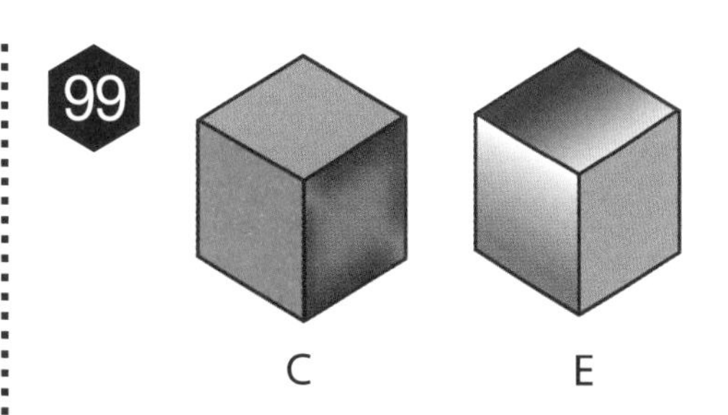

123

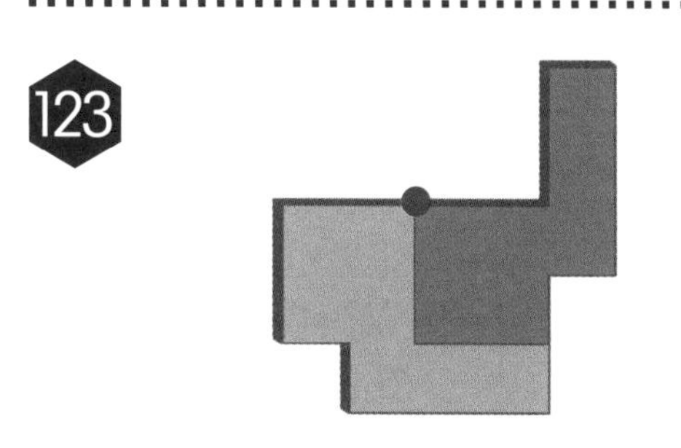

105

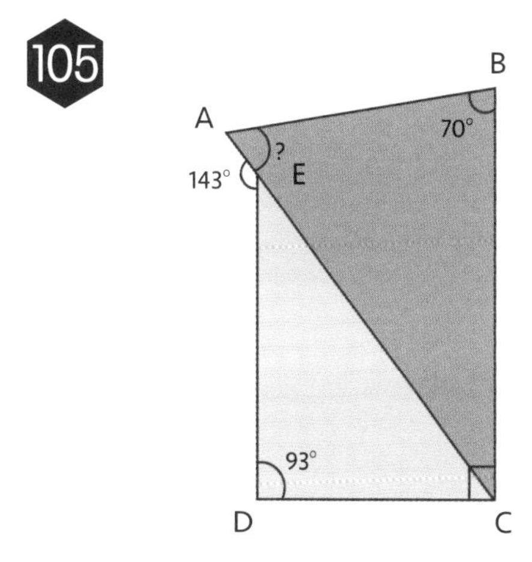

70°.

Add reference points A, B, C and D to the grid. Angle AED is 143°, so angle DEC must be 37°. Angle ECD must be 180° – 37° – 93° = 50°. As angle BCD is a right angle, angle BCA must be 90° - 50° = 40°. Therefore, ? must be 180° – 70° – 40° = 70°.

SOLUTIONS

106

15°.

Each corner of the larger square is 90°, and comprises one angle from an equilateral triangle (60°), and the two smaller angles from an isosceles triangle. These two angles must sum 90° – 60° = 30°. So, the smaller angles of the isosceles triangles must be 30° ÷ 2 = 15°.

107

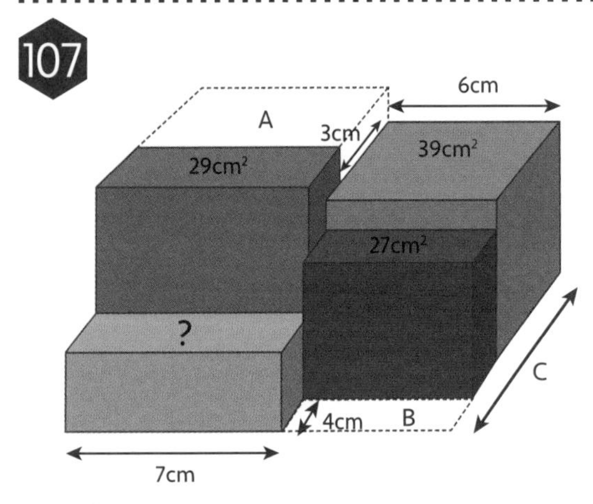

55cm².

Create areas A and B.

A = 7cm x 3cm = 21cm².

B = 6cm x 4cm = 24cm².

C = (24cm² +27cm² + 39cm²) ÷ 6cm = 15cm.

? = (7cm x 15cm) – 21cm2 – 29cm² = 55cm².

131

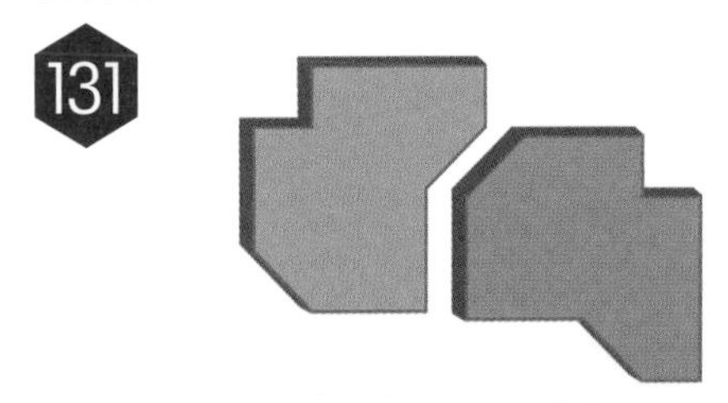

108

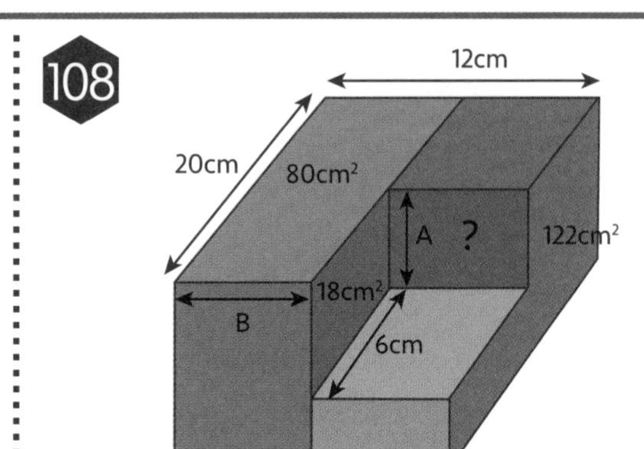

24cm².

A = 18cm² ÷ 6cm = 3cm

B = 80cm² ÷ 20cm = 4cm

? = (12cm - 4cm) x 3cm = 24cm²

109

110

111

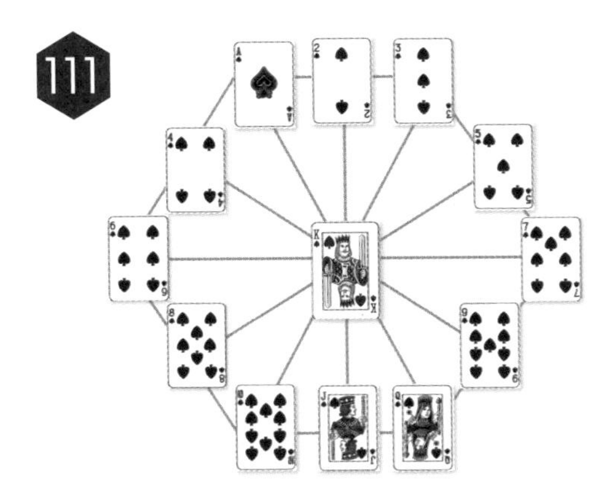

112

113

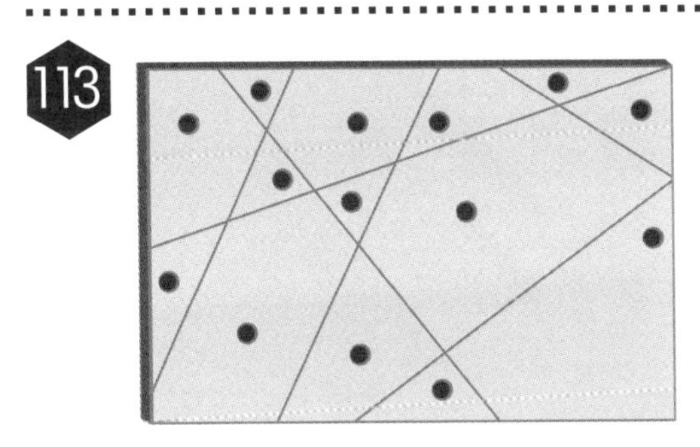

90

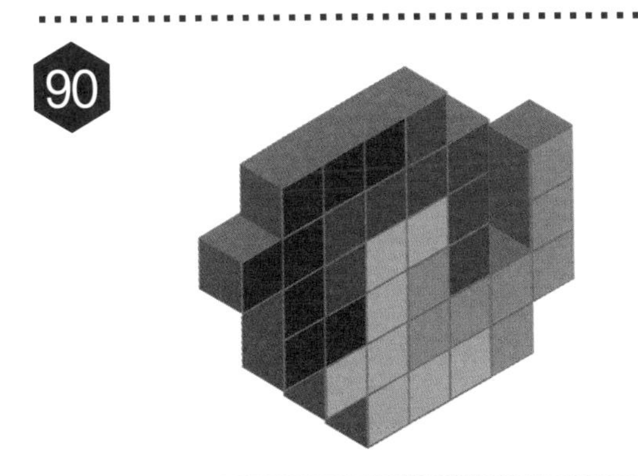

195

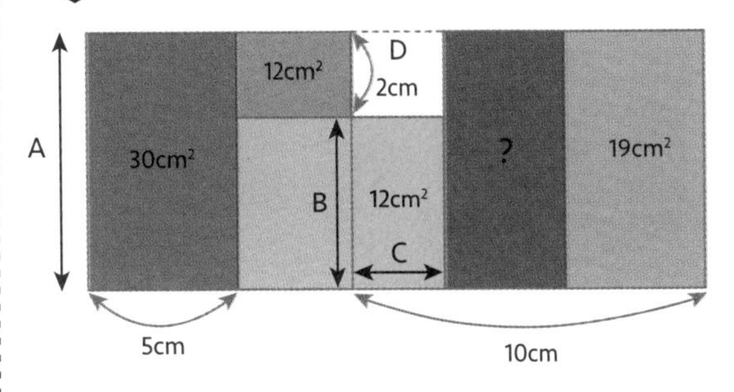

$23cm^2$

A = 6cm. B = 6cm − 2cm = 4cm. C = $12cm^2$ ÷ 4cm = 3cm. Create area D. D = 2cm x 3cm = $6cm^2$. ? = (10cm x (4cm+2cm) − $19cm^2$ − $12cm^2$ − $6cm^2$ = $23cm^2$.

116

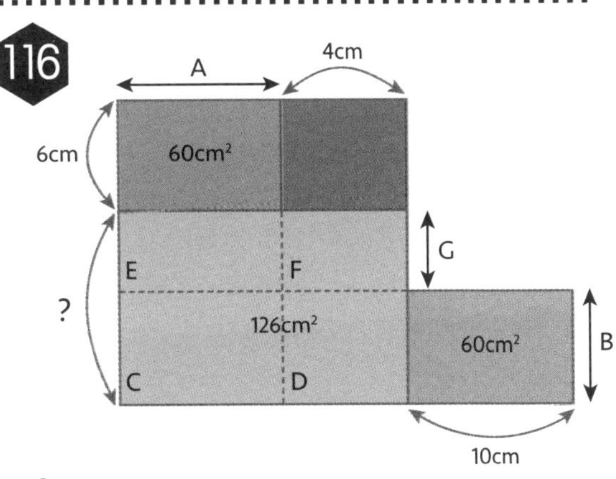

9cm.

A = $60cm^2$ ÷ 6cm = 10cm.

B = $60cm^2$ ÷ 10cm = 6cm.

Create areas C, D, E and F.

C = 10cm x 6cm = $60cm^2$.

D = 4cm x 6cm = $24cm^2$.

E+F = $126cm^2$ − $60cm^2$ − $24cm^2$ = $42cm^2$.

E+F = C+D ÷ 2.

G = 6cm ÷ 2 = 3cm.

? = 6cm + 3cm = 9cm.

SOLUTIONS

76

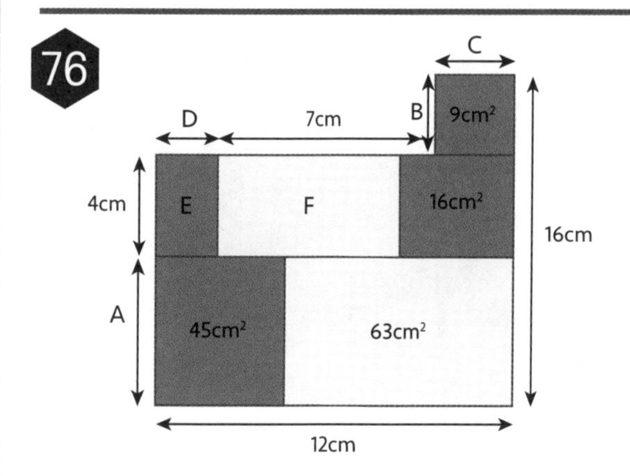

Grey.

A = (45cm^2 + 63cm^2) ÷ 12cm = 9cm.
B = 16cm – 9cm – 4cm = 3cm.
C = 9cm^2 ÷ 3cm = 3cm.
D = 12cm – C – 7cm = 2cm.
E = 4cm x D = 8cm^2.
F = (4cm x 12cm) – E – 16cm^2 = 24cm^2.

118

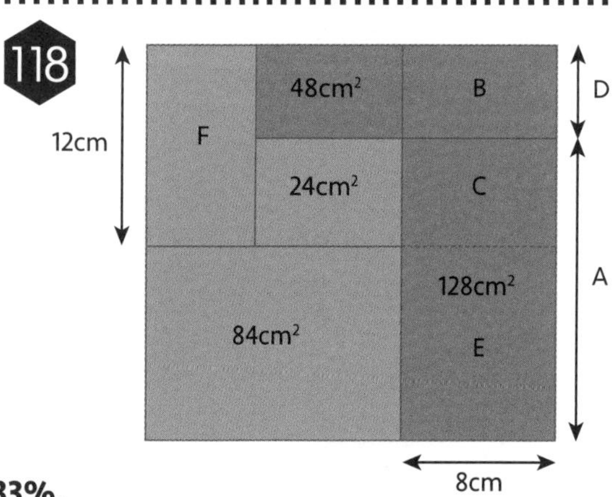

33%.

A = 128cm^2 ÷ 8cm = 16cm. Create area B. 48cm^2 = 24cm^2 x 2, so B = C x 2. Length B + C = 12cm, so D (length B) is 12cm ÷ 3 x 2 = 8cm. B = D x 8cm = 64cm^2, and C = B ÷ 2 = 32cm^2. E = 128cm^2 – C = 96cm^2. E = B + C, so 84cm^2 = 48cm^2 + 24cm^2 + F. F = 84cm^2 – 48cm^2 – 24cm^2 = 12cm^2.

43

120

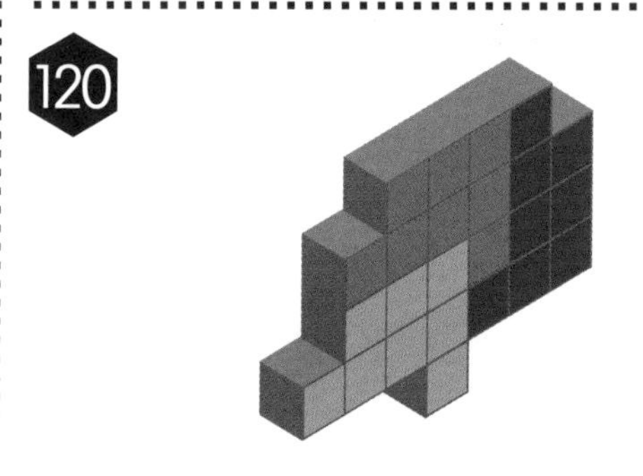

121

81

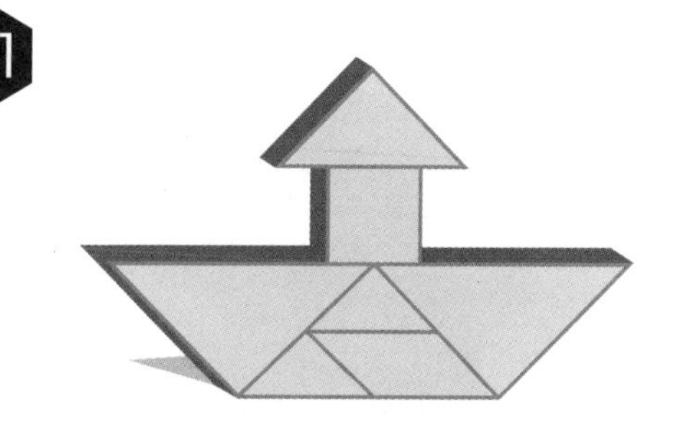

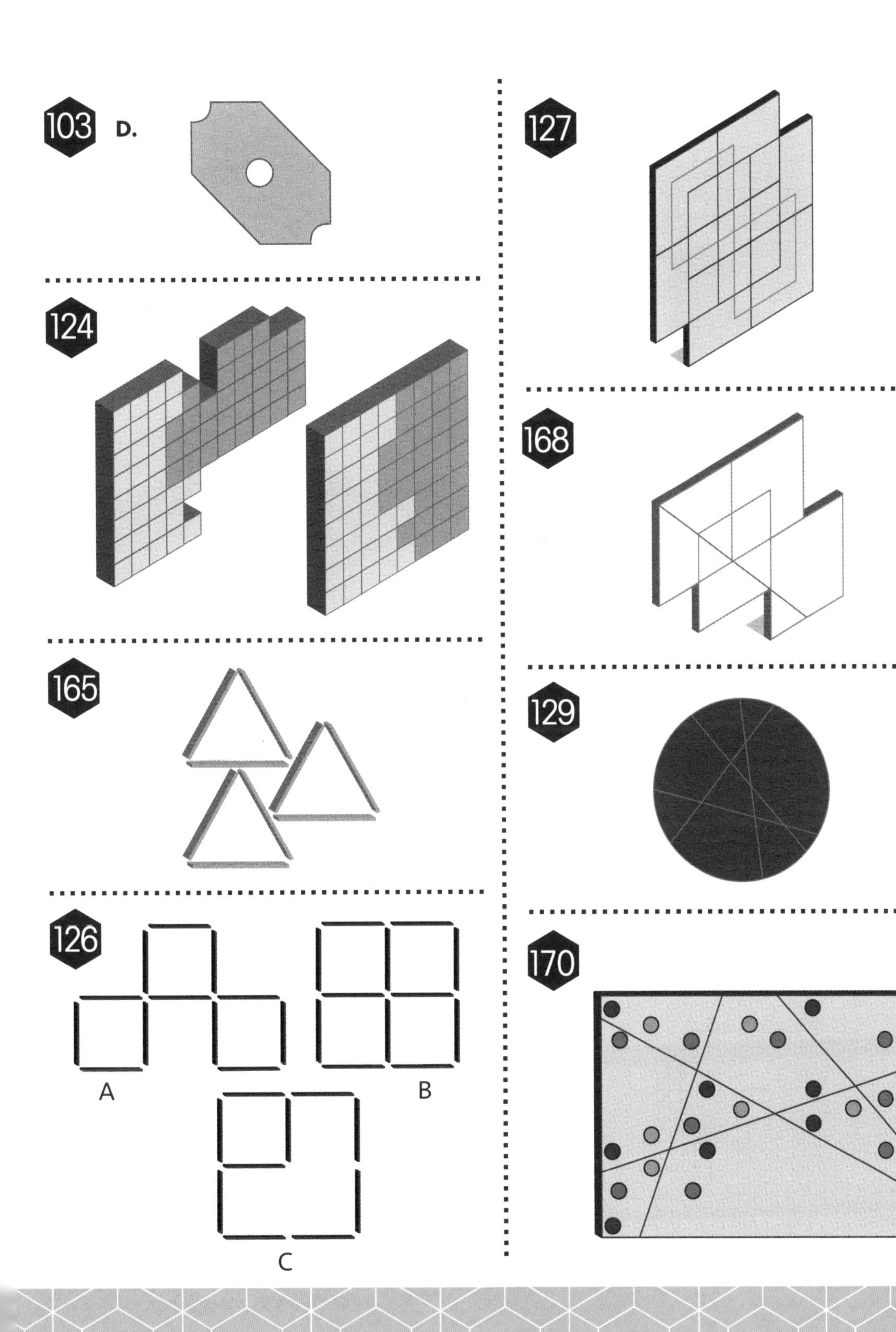
103
D.
124
165
126
A
B
C
127
168
129
170

SOLUTIONS

92cm.

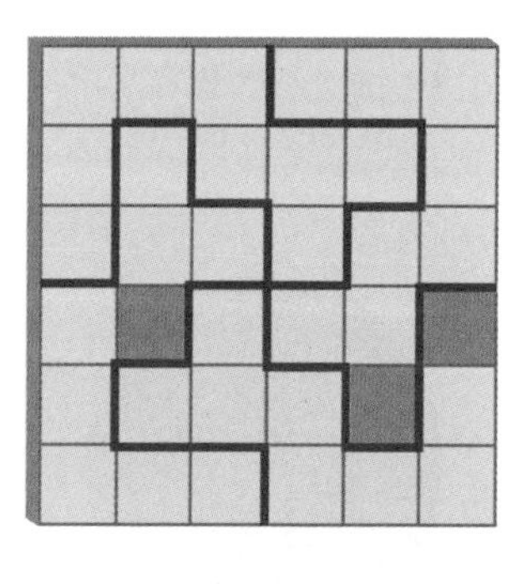

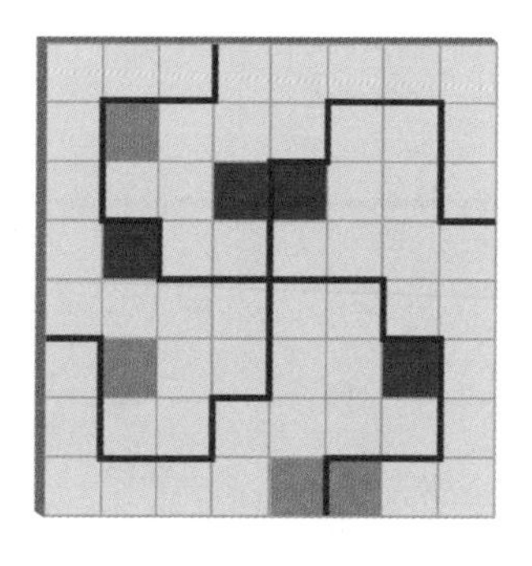

176

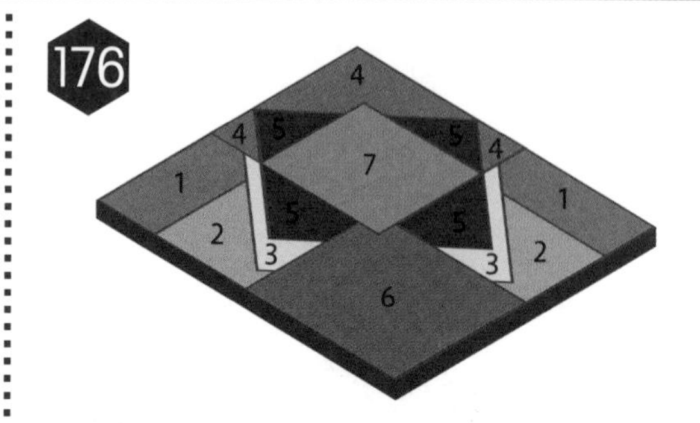

136

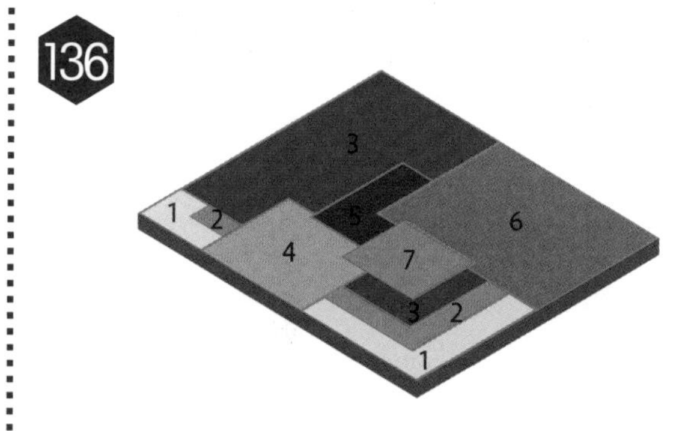

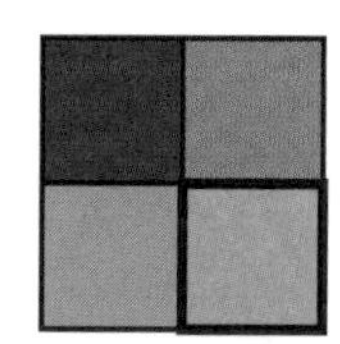

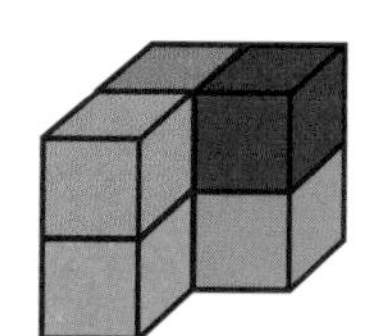

138

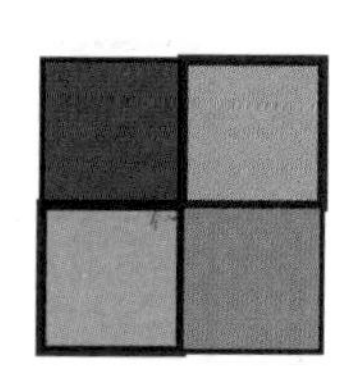

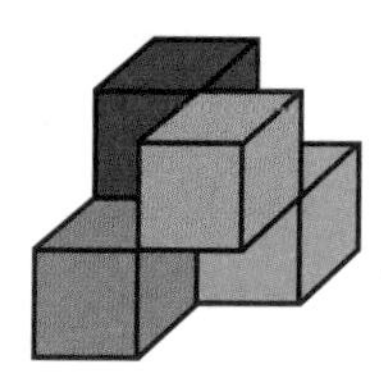

97 **14.**

140 **19.**

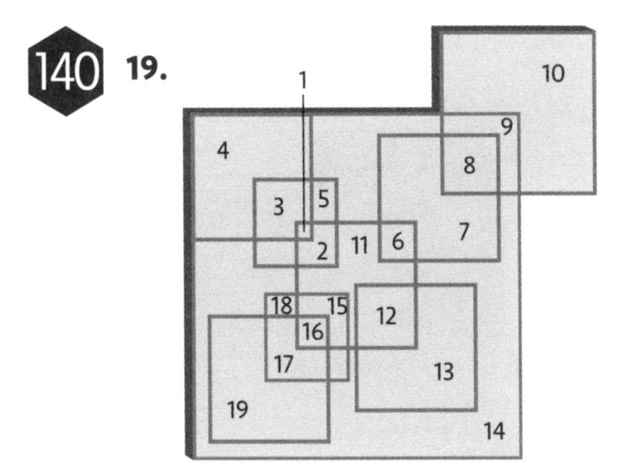

181

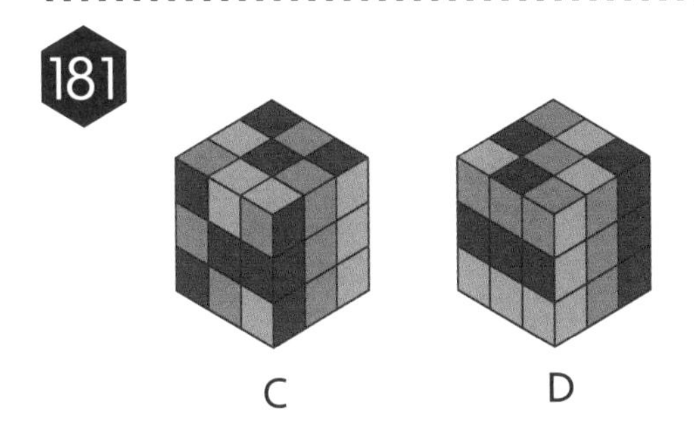

142

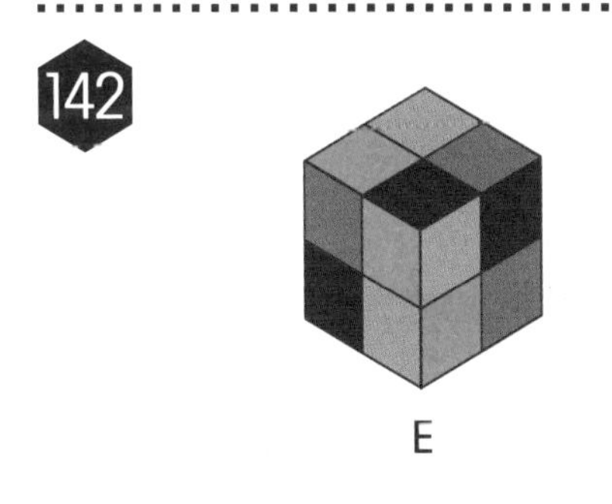

154

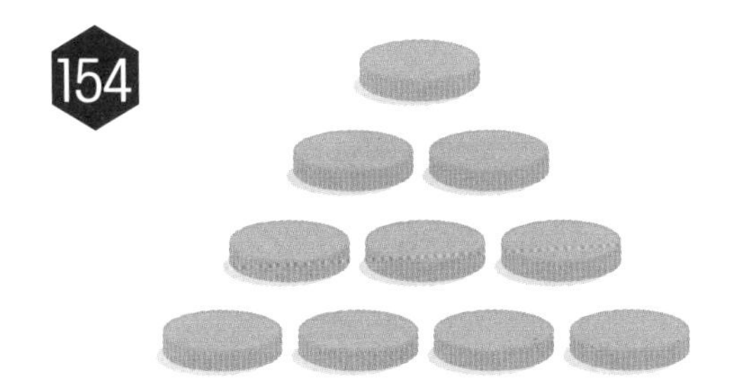

188 **80°.**

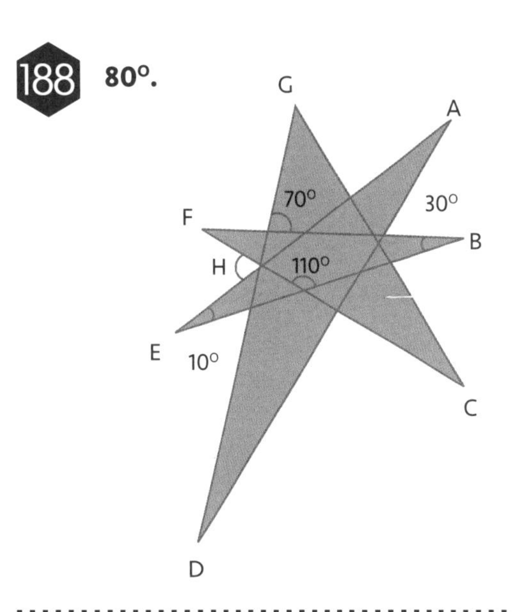

148 **18°.**

As lines AD and EF are parallel, angle CEF = 142°, so angle FEG = 180° – 142° = 38°. Therefore, angle EGF is 180° – 90° – 38° = 52°, and angle GHJ is 180° – 52° – 110° = 18°.

185 **D.**

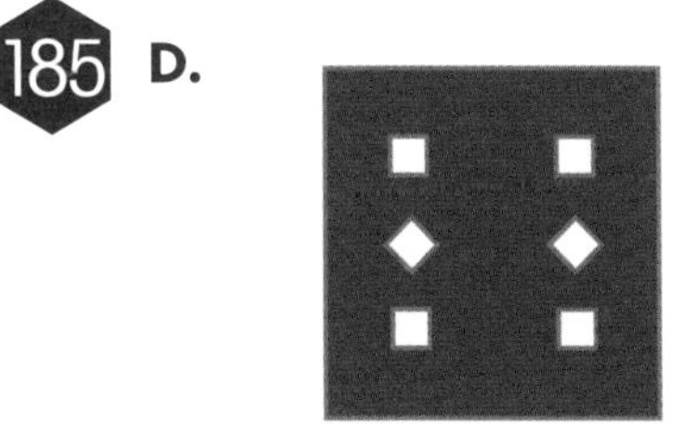

162

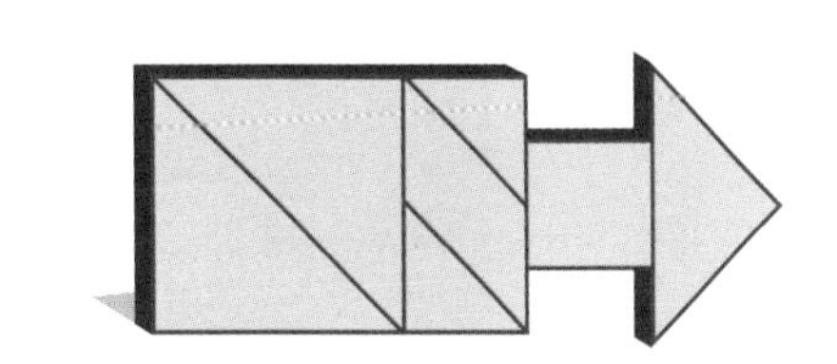

SOLUTIONS

147

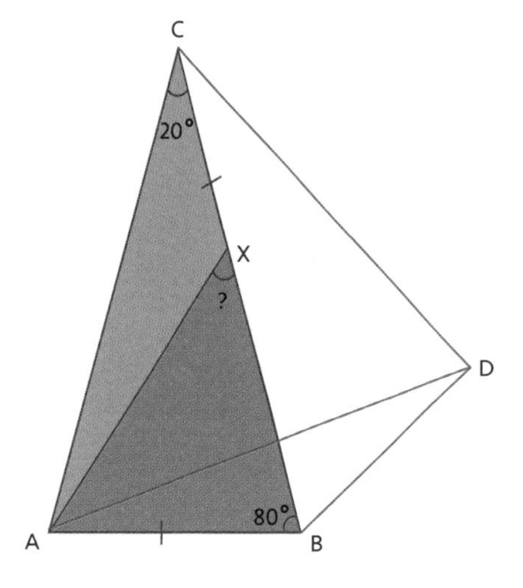

30°.

Angle A must be 180° – 20° – 80° = 80°. This is equal to angle B, so line AC and line BC are equal in length. Next, create an equilateral triangle, ACD. AC=AD=CD. Angle BCD must be 60° – 20° = 40°. Join point B to point D, to create a triangle BCD. Line BC = CD, so CBD and CDB must each be 70°. Angle CDA is 60°, so angle ADB must be 10°. and thus angle DAB 20°. As line AB = line CX, and line AC = line AD and angle ACX = angle DAB then triangles ACX and DAB are congruent (side, angle, side). Therefore, angle AXC must correspond with angle ABD (150°) and ? = 180° – 150° = 30°.

145 **B.**

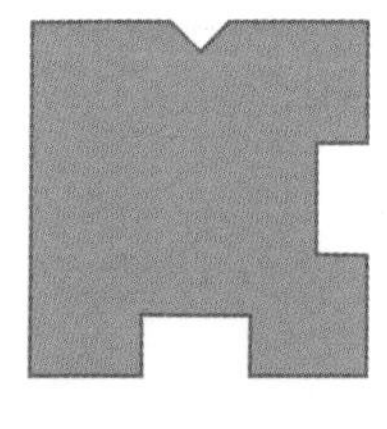

149

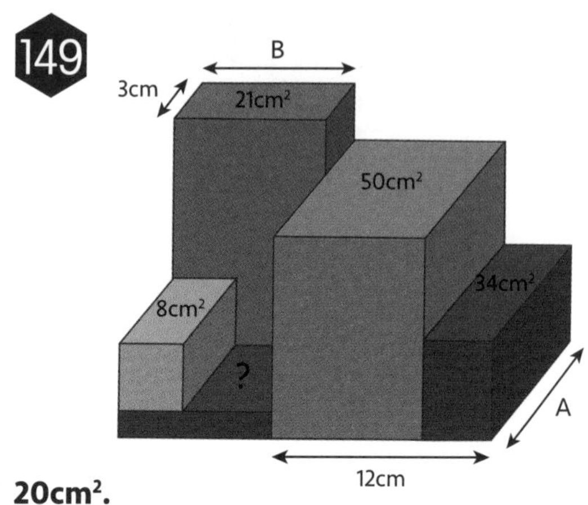

20cm².

A = (50cm² + 34cm²) ÷ 12cm = 7cm.

B = 21cm² ÷ 3cm = 7cm.

? = A x B – (21cm² + 8cm²) = 20cm².

189

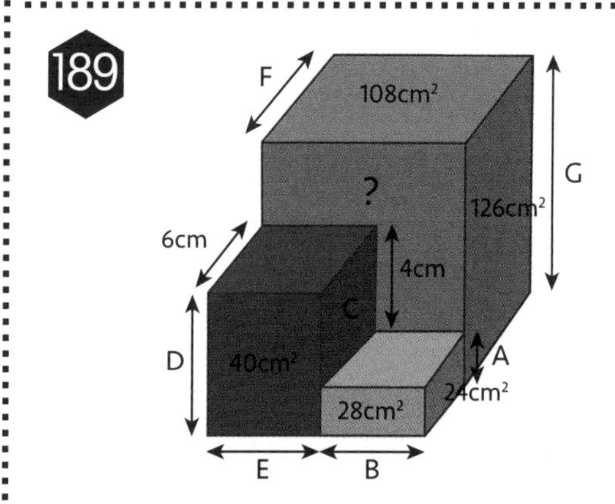

100cm.

A = 24cm² ÷ 6cm = 4cm.

B = 28cm² ÷ 4cm = 7cm.

C = 6cm x 4cm = 24cm².

D = 4cm + 4cm = 8cm.

E = 40cm² ÷ 8cm = 5cm.

F = 108cm² ÷ (E + B) = 9cm.

G = 126cm² ÷ 9cm = 14cm.

? = G x (E+B) – 40cm² – 28cm² = 100cm².

191

152

164

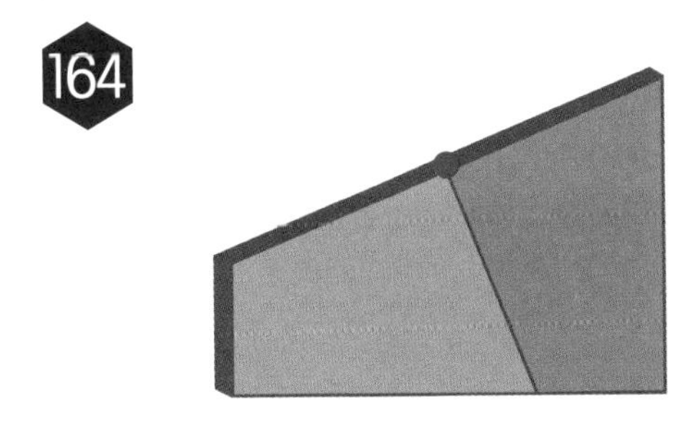

143

9cm.

The perimeter is equivalent to 54 widths.

486cm ÷ 54 = 9cm

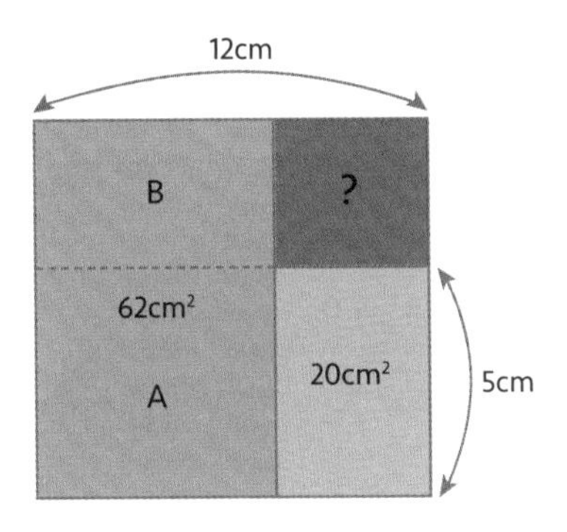

$11cm^2$.

Create areas A and B.

A = 5cm x 12cm − $20cm^2$ = $40cm^2$.

$20cm^2$ = $40cm^2$ ÷ 2.

B = $62cm^2$ − $40cm^2$ = $22cm^2$.

? = $22cm^2$ ÷ 2 = $11cm^2$.

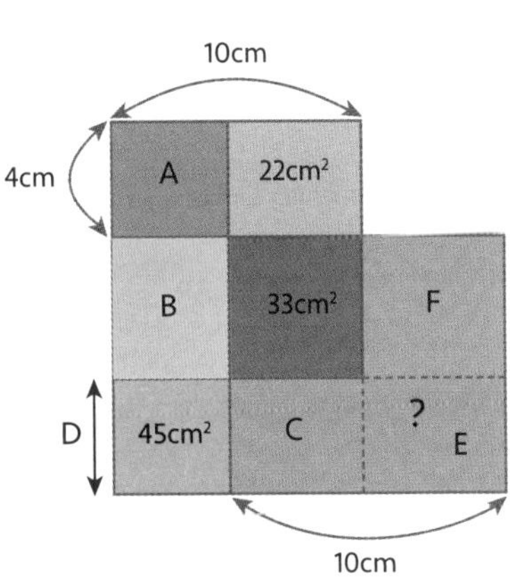

$127cm^2$.

A = (10cm x 4cm) − $22cm^2$ = $18cm^2$. Create area B. $33cm^2$ = $22cm^2$ ÷ 2 x 3, so B = $18cm^2$ ÷ 2 x 3 = $27cm^2$. Create area C. A + B = $45cm^2$, so C = $22cm^2$ + $33cm^2$ = $55cm^2$. C + $45cm^2$ = $100cm^2$, so D = $100cm^2$ ÷ 10cm = 10cm. Create areas E and F. As the length of C + E is 10cm, E = $45cm^2$ and F = B = $27cm^2$. ? = $55cm^2$ + $45cm^2$ + $27cm^2$ = $127cm^2$.

SOLUTIONS

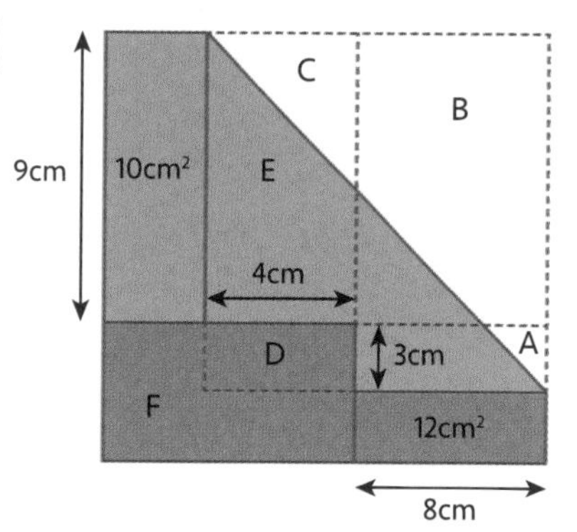

One third. (35/105).

Create areas A, B, C, D.

A = 3cm x 8cm = 24cm^2.

B = 9cm x 8cm = 72cm^2.

C = 4cm x 9cm = 36cm^2.

D = 4cm x 3cm = 12cm^2.

A+B+C+D = 144cm^2.

The irregular polygon E = 144cm^2 ÷ 2 – D = 60cm^2.

12cm^2 + A = B ÷ 2, so D+F = (10cm^2 + C) ÷ 2 = 23cm^2.

12.

200

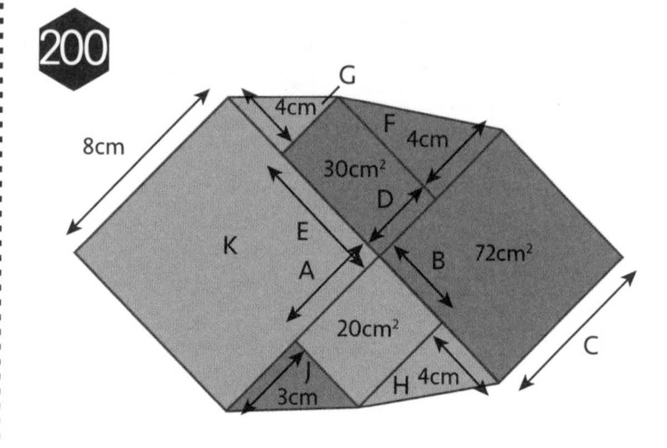

50%.

A = 8cm – 3cm = 5cm. B = 20cm^2 ÷ 5cm = 4cm. C = 72cm^2 ÷ 8cm = 9cm. D = 9cm – 4cm = 5cm. E = 30cm^2 ÷ 5cm = 6cm. F = (6cm x 4cm) ÷ 2 = 12cm^2. G = (D x 4cm) ÷ 2 = 10cm^2. H = (A x 4cm) ÷ 2 = 10cm^2. J = (B x 3cm) ÷ 2 = 6cm^2. K = 8cm x (4cm + E) = 80cm^2.

160

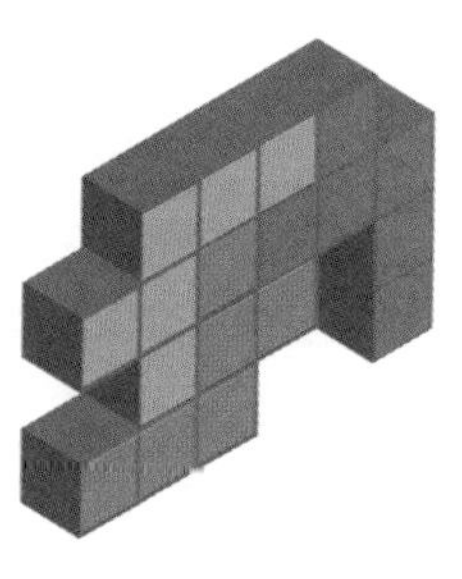

161

163

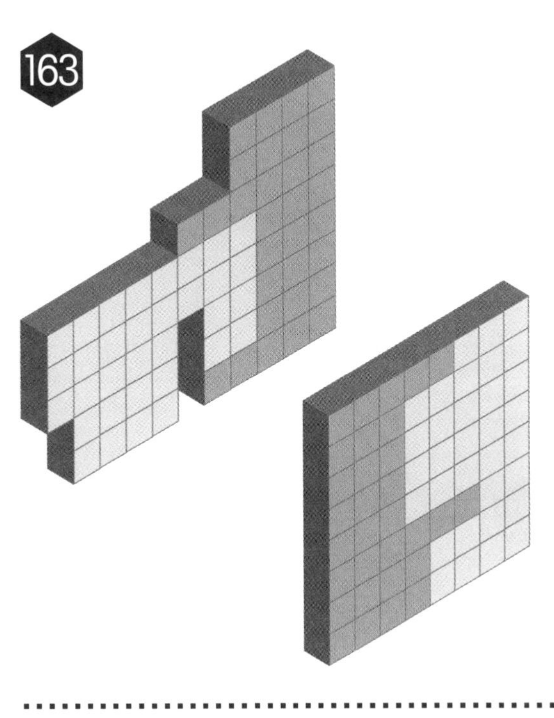

193

86

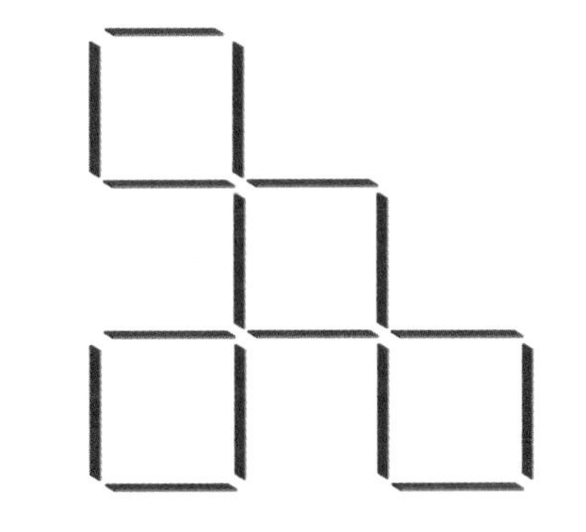

166

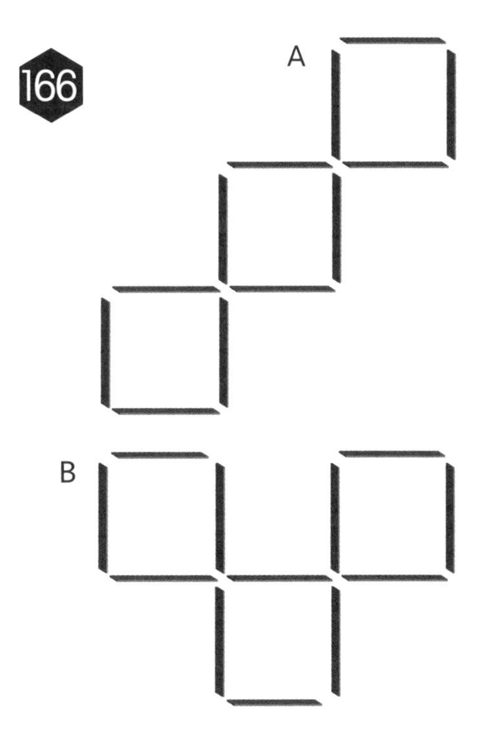

167

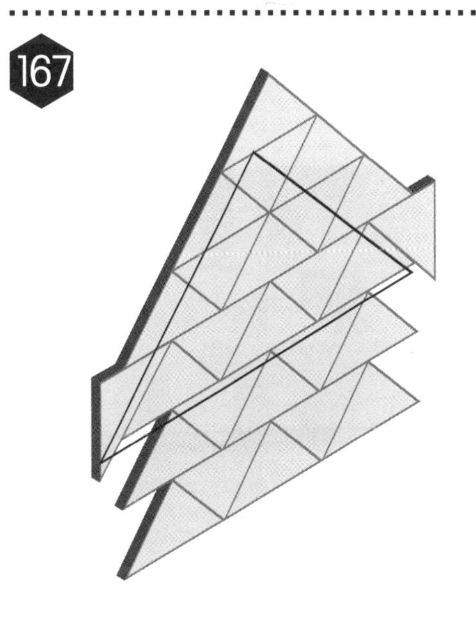

SOLUTIONS

128

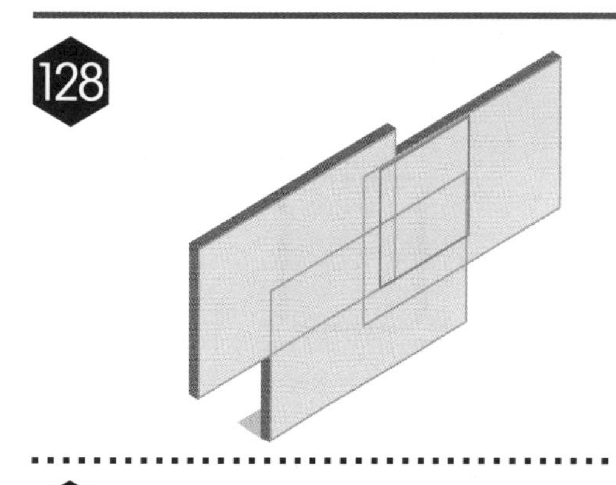

169

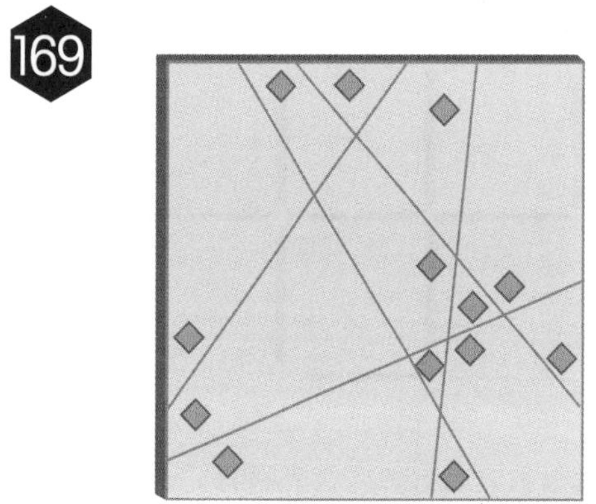

151

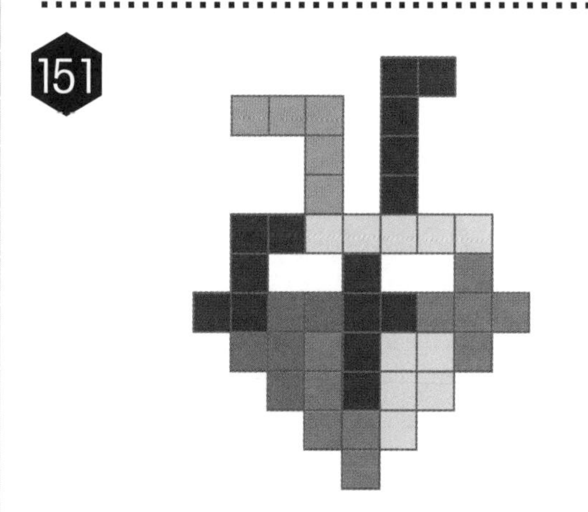

171

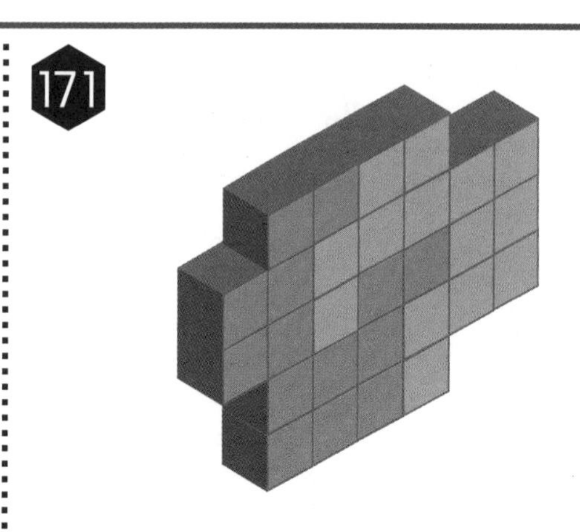

172

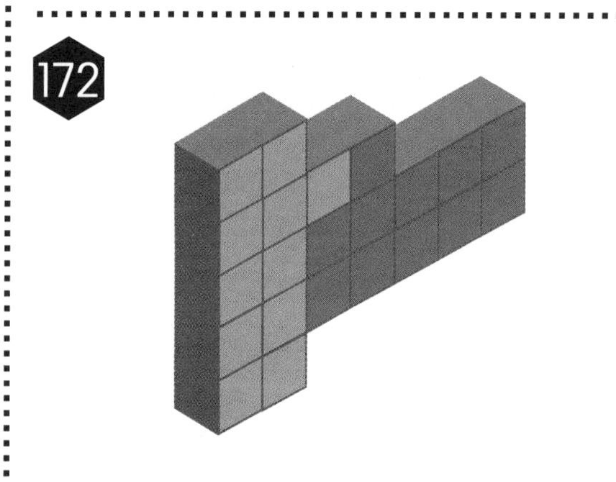

173

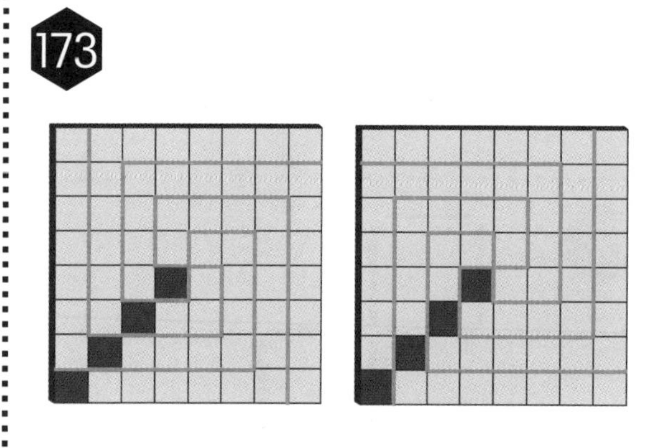

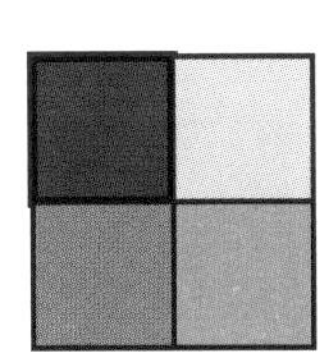

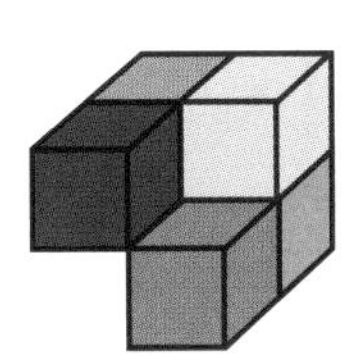

33.

175

135

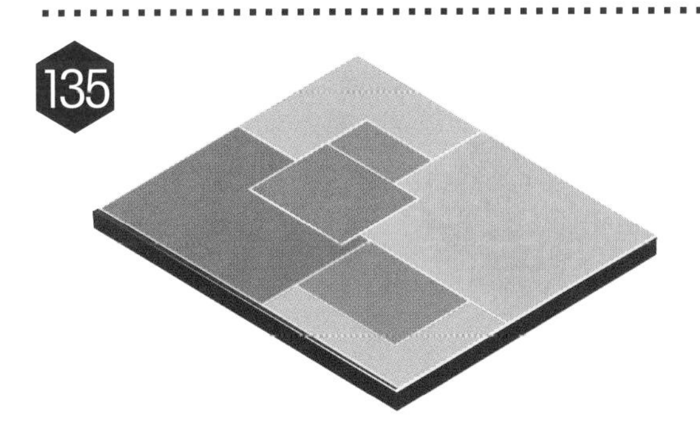

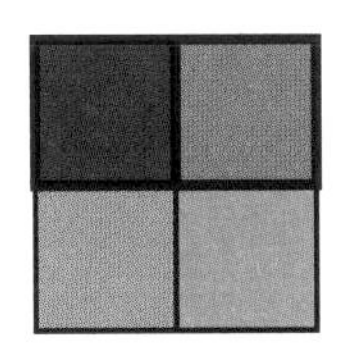

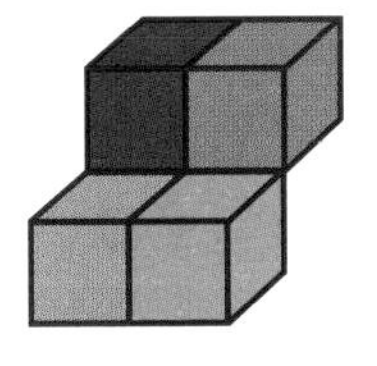

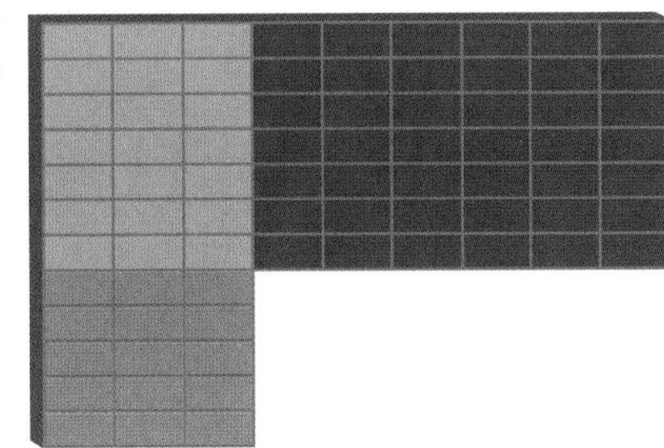

1560.

Firstly, work out the rectangles in the yellow and red areas combined:
((vertical lines x (vertical lines - 1)) x ((horizontal lines x (horizontal lines – 1)) ÷ 4 = 1260.
Next, work out the rectangles in the blue and yellow areas combined:
((13x12) x (4x3)) ÷ 4 = (156 x 12) ÷ 4 = 468.
Add these two together: 1728.
Then, work out the recrangles in the yellow area, which have already been counted in the yellow and blue total: ((8x7) x (4x3)) ÷ 4 = 168.

The total number of rectangles is
1728 – 168 = 1560.

SOLUTIONS

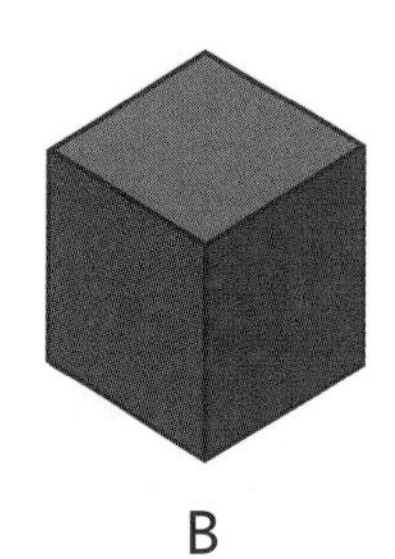

B

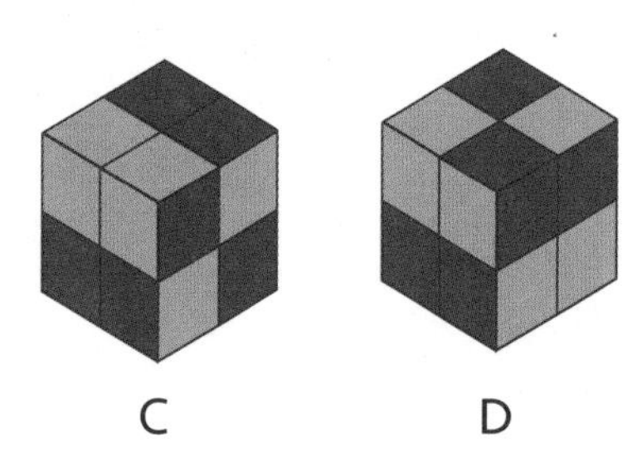

C D

36cm, 252cm.

The square tile must have sides of 12cm, so the rectangular tile must have sides of 6cm and 12cm. The overall perimeter is the equivalent of 42 widths, or 21 heights = 252cm.

1200cm.

The perimeter of the wall, including holes is 48 lengths and 48 heights of brick, which is the equivalent to the perimeter of 24 bricks.

146 **C.**

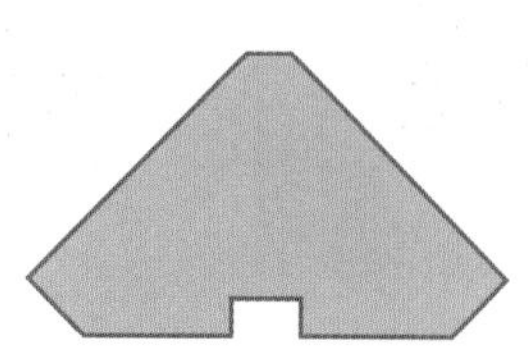

186 **B.**

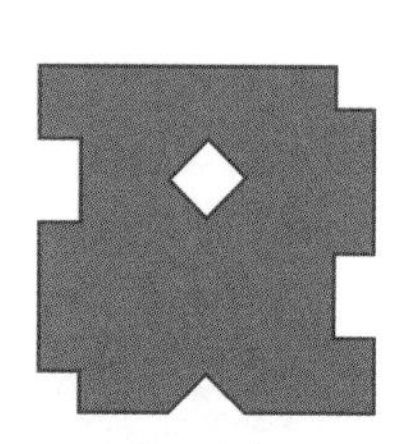

187

40°.

For ease of reference, name each angle on the perimeter, A to E, P to T, as shown. Angle APT must be 70° (sum of angles on a straight line). Similarly, angle TSE must be 60°, and angle ETS must be 70° (sum of internal angles of a triangle). Angle ATP is vertically opposite angle ETS, so must equal 70°. So, angle TAP must equal 180° – 70° – 70° = 40°.

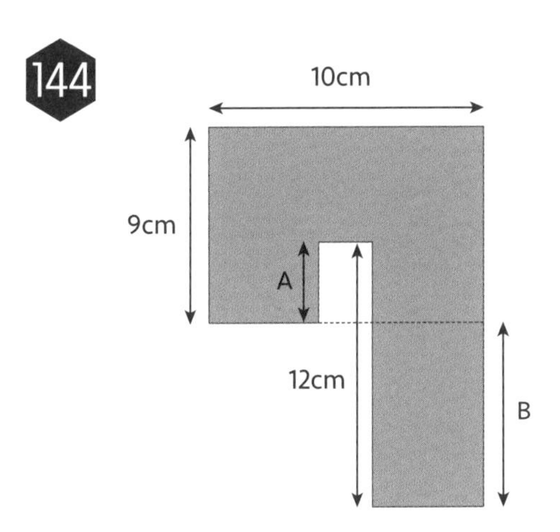

62cm.

Length A + Length B = 12cm.

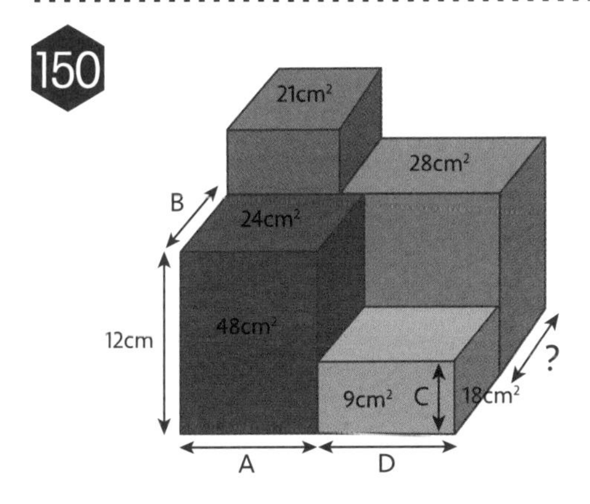

7cm.

A = 48cm² ÷ 12cm = 4cm.

B = 24cm² ÷ 4cm = 6cm.

C = 18cm² ÷ 6cm = 3cm.

D = 9cm² ÷ 3cm = 3cm.

A + D = 7cm

? = (21cm² + 28cm²) ÷ 7cm = 7cm

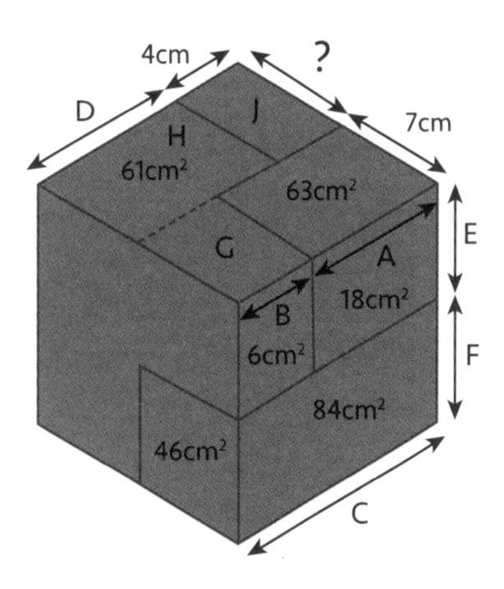

5cm

A = 63cm² ÷ 7cm = 9cm.

B = 6cm² ÷ 2cm = 3cm.

C = A + B = 12cm.

D = C – 4cm = 8cm.

E = 18cm² ÷ 9cm = 2cm .

F = 84cm² ÷ C = 7cm.

Create areas G and H.

G = 7cm x 3cm = 21cm².

H = 61cm² – 21cm² = 40cm².

As H is twice as long as area J, area J must be 20cm², so ? = 20cm² ÷ 4cm = 5cm.

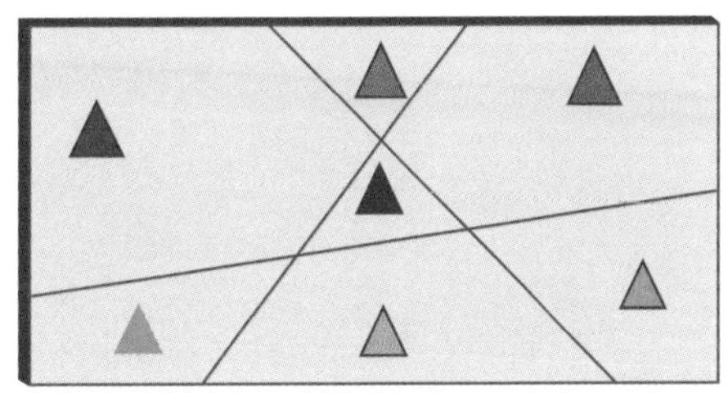

SOLUTIONS

192

153

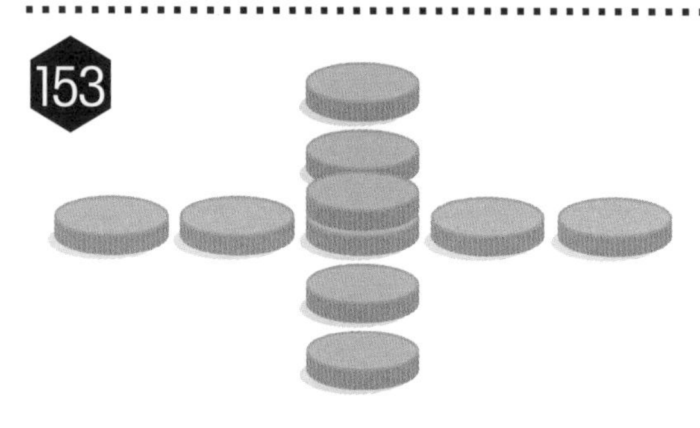

Move one of the end discs from the six-coin row and place it on top of the centre disc of the five-disc row.

194

155

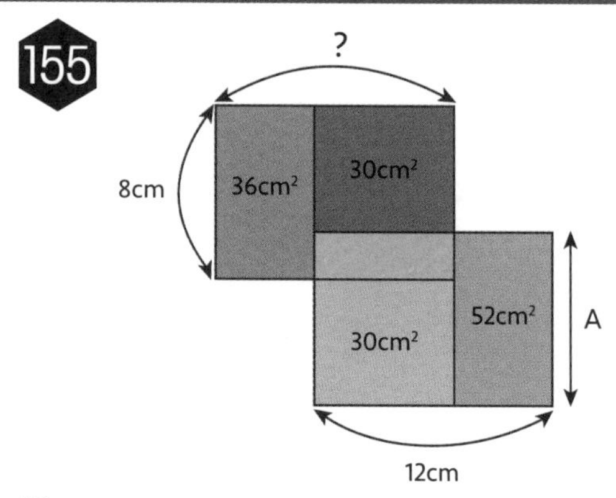

10cm.

A = 8cm.

B = (8cm x 12cm) – $52cm^2$ – $30cm^2$ = $14cm^2$.

? = $36cm^2$ + $30cm^2$ + $14cm^2$) ÷ 8cm = 10cm.

196

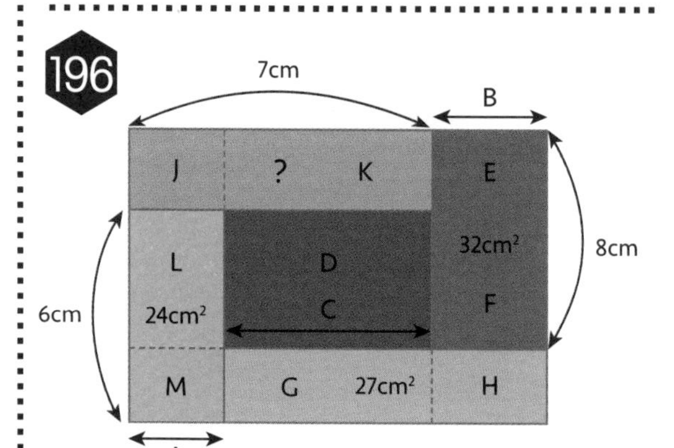

$41cm^2$.

A = $24cm^2$ ÷ 6cm = 4cm. B = $32cm^2$ ÷ 8cm = 4cm. C = 7cm – A (4cm) = 3cm. Mark area D, create areas E, F, G, H, J, K, L and M. D + K = (3cm + 4cm) x 8cm – $32cm^2$ = $24cm^2$.

D + F = 6cm x 7cm – $27cm^2$ = $15cm^2$.

E + K = 7cm x 8cm – $15cm^2$ = $41cm^2$.

As E has the same dimesnions as J, J + K = $41cm^2$. ? = 41 cm^2.

197

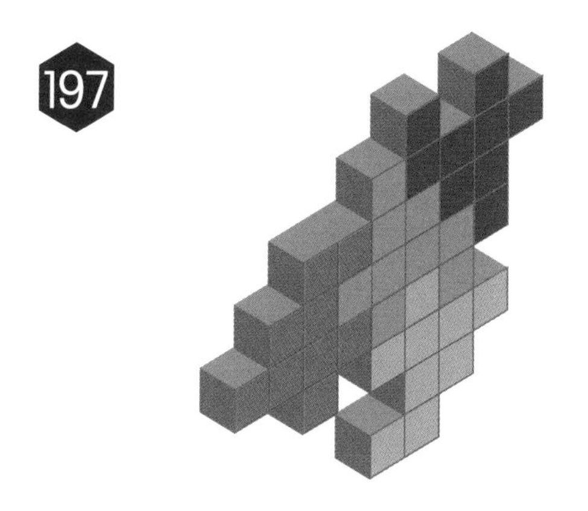

198

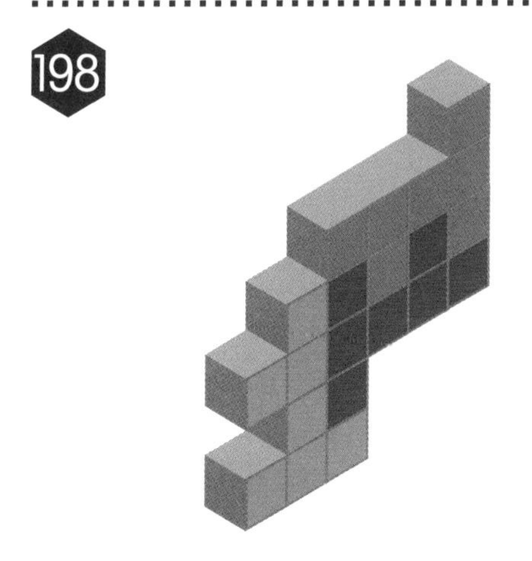

199

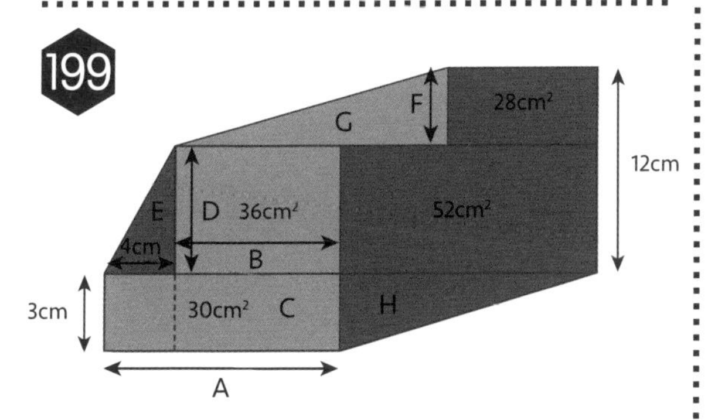

Red 105cm² Blue 96cm².

A = 30cm² ÷ 3cm = 10cm.

B = 10cm − 4cm = 6cm.

C = 6cm x 3cm = 18cm².

D = 36cm² ÷ 6cm = 6cm.

E = (4cm x 6cm) ÷ 2 = 12cm².

F = 12cm − 6cm = 6cm.

G x 2 + 28cm² = 36cm² + 52cm² = 88cm².

G = (88cm² − 28cm²) ÷ 2 = 30cm².

H x 2 = 52cm² ÷ 2 = 26cm².

H = 26cm² ÷ 2 = 13cm².

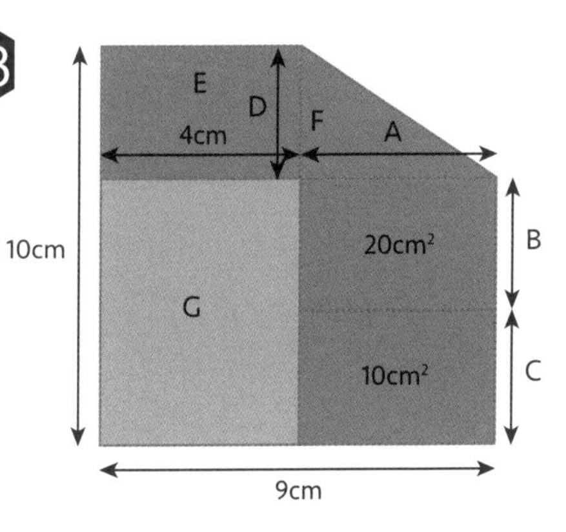

30%

A = 9cm − 4cm = 5cm.

B = 20cm² ÷ 5cm = 4cm.

C = 10cm² ÷ 5cm = 2cm.

D = 10cm − B − C = 4cm.

E = D x 4cm = 16cm².

F = (A x D) ÷ 2 = 10cm².

G = 4cm x (B + C) = 24cm².

NOTES